METHODS IN MOLECULAR BIOLOGY

Series Editor
John M. Walker
School of Life and Medical Sciences
University of Hertfordshire
Hatfield, Hertfordshire, AL10 9AB, UK

For further volumes:
http://www.springer.com/series/7651

The Renin-Angiotensin-Aldosterone System

Methods and Protocols

Edited by

Sean E. Thatcher

Department of Pharmacology and Nutritional Sciences, University of Kentucky, Lexington, KY, USA

Humana Press

Editor
Sean E. Thatcher
Department of Pharmacology and Nutritional Sciences
University of Kentucky
Lexington, KY, USA

ISSN 1064-3745 ISSN 1940-6029 (electronic)
Methods in Molecular Biology
ISBN 978-1-4939-7028-5 ISBN 978-1-4939-7030-8 (eBook)
DOI 10.1007/978-1-4939-7030-8

Library of Congress Control Number: 2017937363

Printed on acid-free paper

This Humana Press imprint is published by Springer Nature
The registered company is Springer Science+Business Media LLC
The registered company address is: 233 Spring Street, New York, NY 10013, U.S.A.

Preface

The purpose of this book is to provide scientists with detailed protocols that will help in measuring different components of the renin-angiotensin-aldosterone system (RAAS). This book also helps in the use of new methods to measure angiotensin peptides and to apply cutting-edge techniques to discern the influence of RAAS components on different aspects of mammalian disease. Each chapter provides an in-depth focus on each experimental technique and gives the reader the best approach to examine how the RAAS might influence his/her own experimental outcomes.

Lexington, KY, USA *Sean E. Thatcher*

Contents

Contributors

MAIIA E. BRAGINA • *Laboratory of Hemodynamics and Cardiovascular Technology, Institute of Bioengineering, Ecole Polytechnique Fédérale de Lausanne, Lausanne, Switzerland*

LISA A. CASSIS • *Department of Pharmacology and Nutritional Sciences, University of Kentucky, Lexington, KY, USA*

MARK C. CHAPPELL • *Hypertension and Vascular Research, Department of Surgery, Cardiovascular Scineces Center, Winston-Salem, NC, USA*

STEFAN CHLOPICKI • *Jagiellonian Centre for Experimental Therapeutics (JCET), Jagiellonian University, Krakow, Poland; Department of Experimental Pharmacology, Jagiellonian University Medical College, Krakow, Poland*

STEVEN D. CROWLEY • *Division of Nephrology, Department of Medicine, Durham VA and Duke University Medical Centers, Durham, NC, USA*

ALAN DAUGHERTY • *Saha Cardiovascular Research Center, University of Kentucky, Lexington, KY, USA; Department of Physiology, University of Kentucky, Lexington, KY, USA*

SATORU EGUCHI • *Department of Physiology and Cardiovascular Research Center, Lewis Katz School of Medicine, Temple University, Philadelphia, PA, USA*

KHALID M. ELASED • *Department of Pharmacology and Toxicology, Boonshoft School of Medicine, Wright State University, Dayton, OH, USA*

KATHERINE J. ELLIOTT • *Department of Physiology and Cardiovascular Research Center, Lewis Katz School of Medicine, Temple University, Philadelphia, PA, USA*

MING C. GONG • *Department of Physiology, University of Kentucky, Lexington, KY, USA; Saha Cardiovascular Research Center, University of Kentucky, Lexington, KY, USA*

JUSTIN L. GROBE • *Department of Pharmacology, Center for Hypertension Research, The Obesity Research and Education Initiative, François M. Abboud Cardiovascular Research Center, The Fraternal Order of Eagles' Diabetes Research Center, University of Iowa, Iowa City, IA, USA*

NADJA GROBE • *Department of Pharmacology and Toxicology, Boonshoft School of Medicine, Wright State University, Dayton, OH, USA*

ZHENHENG GUO • *Department of Physiology, University of Kentucky, Lexington, KY, USA; Research and Development, Lexington Veterans Affairs Medical Center, University of Kentucky, Lexington, KY, USA; Saha Cardiovascular Research Center, University of Kentucky, Lexington, KY, USA; Department of Pharmacology and Nutritional Science, University of Kentucky, Lexington, KY, USA*

YAGNA P.R. JARAJAPU • *Department of Pharmaceutical Sciences, College of Health Professions, North Dakota State University, Fargo, ND, USA*

SHU LIU • *Department of Physiology, University of Kentucky, Lexington, KY, USA; Research and Development, Lexington Veterans Affairs Medical Center, University of Kentucky, Lexington, KY, USA; Saha Cardiovascular Research Center, University of Kentucky, Lexington, KY, USA*

HONG LU • *Department of Physiology, University of Kentucky, Lexington, KY, USA; Saha Cardiovascular Research Center, University of Kentucky, Lexington, KY, USA*

MARIOLA OLKOWICZ • *Department of Biochemistry, Medical University of Gdansk, Gdansk, Poland; Department of Biotechnology and Food Microbiology, Poznan University of Life Sciences, Poznan, Poland*
NATHAN P. RUDEMILLER • *Division of Nephrology, Department of Medicine, Durham VA and Duke University Medical Centers, Durham, NC, USA*
RODRIGO A. FRAGA-SILVA • *Laboratory of Hemodynamics and Cardiovascular Technology, Institute of Bioengineering, Ecole Polytechnique Fédérale de Lausanne, Lausanne, Switzerland*
RYSZARD T. SMOLENSKI • *Department of Biochemistry, Medical University of Gdansk, Gdansk, Poland*
NIKOLAOS STERGIOPULOS • *Laboratory of Hemodynamics and Cardiovascular Technology, Institute of Bioengineering, Ecole Polytechnique Fédérale de Lausanne, Lausanne, Switzerland*
SEAN E. THATCHER • *Department of Pharmacology and Nutritional Sciences, University of Kentucky, Lexington, KY, USA*
RHIAN M. TOUYZ • *Institute of Cardiovascular and Medical Sciences, BHF Glasgow Cardiovascular Research Centre, University of Glasgow, Glasgow, UK*
SOFIA TSIROPOULOU • *Institute of Cardiovascular and Medical Sciences, BHF Glasgow Cardiovascular Research Centre, University of Glasgow, Glasgow, UK*
YU WANG • *Department of Pharmacology and Nutritional Sciences, University of Kentucky, Lexington, KY, USA*
BRYAN A. WILSON • *McAllister Heart Institute, University of North Caroliina Chapel Hill, Winston-Salem, NC, USA*
CONGQING WU • *Saha Cardiovascular Research Center, University of Kentucky, Lexington, KY, USA*

Chapter 1

A Brief Introduction into the Renin-Angiotensin-Aldosterone System: New and Old Techniques

Sean E. Thatcher

Abstract

The renin-angiotensin-aldosterone system (RAAS) is a complex system of enzymes, receptors, and peptides that help to control blood pressure and fluid homeostasis. Techniques in studying the RAAS can be difficult due to such factors as peptide/enzyme stability and receptor localization. This paper gives a brief account of the different components of the RAAS and current methods in measuring each component. There is also a discussion of different methods in measuring stem and immune cells by flow cytometry, hypertension, atherosclerosis, oxidative stress, energy balance, and other RAAS-activated phenotypes. While studies on the RAAS have been performed for over 100 years, new techniques have allowed scientists to come up with new insights into this system. These techniques are detailed in this Methods in Molecular Biology Series and give students new to studying the RAAS the proper controls and technical details needed to perform each procedure.

Key words Angiotensin, Historical, RAAS, Techniques, Methods

1 Introduction

The renin-angiotensin-aldosterone system (RAAS) is an important endocrine system responsible for the control of blood pressure, sodium levels, and extracellular fluid homeostasis. Since the discovery of renin, by Tiegerstedt and Bergman in 1898, a tremendous amount of work has been published in understanding the RAAS and the involvement of this system in pathophysiology. There are numerous reviews on the RAAS, and this review will focus on components of the system and assays that are now typically used in the study of the RAAS.

1.1 Renin

Renin was the first component to be measured by Tiegerstedt and Bergman when they injected kidney extracts into rabbits and noticed that this induced hypertension in their model [1]. Since this time, work on purification of renin (40 kilo-dalton (kDa)) and understanding the role of renin activity in plasma of different animal models has

Sean E. Thatcher (ed.), *The Renin-Angiotensin-Aldosterone System: Methods and Protocols*, Methods in Molecular Biology, vol. 1614, DOI 10.1007/978-1-4939-7030-8_1, © Springer Science+Business Media LLC 2017

helped to discern the overall activation of the RAAS. Renin is an aspartate protease that cleaves angiotensinogen to the decapeptide, angiotensin I (AngI). It has been shown that low salt or use of an angiotensin 1 receptor (AT1R) blocker/ACE inhibitor (ACEi) can increase renin release [2, 3]. Renin release can also be induced by anesthetics, such as pentobarbital, so it is important to be aware of the specific drugs used as euthanasia for experimental models [4]. It has also been shown that giving exogenous angiotensin II (AngII) can inhibit renin release through a negative feedback mechanism [5]. While some measures of renin could include mRNA or protein via real-time PCR or immunoblotting respectively, the best measurement of renin is through renin activity. Typically, blood is drawn from the animal with an inhibitor cocktail that contains ethylene diaminotetracetic acid (EDTA), and other inhibitors of RAAS components for aminopeptidases, endopeptidases, and serine/cysteine proteases. The blood is then spun by centrifugation and the plasma is measured for renin activity. For mice, this is typically done through measurement of AngI, before and after giving exogenous angiotensinogen, typically from another source, such as rat. After a period of time, the reaction is terminated by incubating at a boiling temperature and then the samples are taken through a radioimmunoassay (RIA) for AngI. Some will report plasma renin concentrations (PRC) as the difference of AngI levels before and after giving exogenous angiotensinogen. However, this is not the best method [4]. PRCs would be the AngI levels before giving any exogenous substrate. These concentrations would be different under the above conditions, such as a salt diet. For plasma renin activity (PRA), this would be the value of AngI formed under exogenous angiotensinogen at 37 °C for 30 min to 1 h of incubation. The PRA value will then allow the determination if renin activity is altered under different experimental conditions. Typically, results for PRA are given as nanograms/milliliter/hour (ng/mL/h). It is important to note that performing this assay requires the use of both positive and negative controls to ensure the accuracy of the measurement. An example of a positive control would be the use of an AT1R blocker in an animal to get a higher renin measurement. An example of a negative control could be the measurement of the sample with a renin inhibitor present, such as aliskiren. It is also important to note that while PRA can be used to measure systemic renin activity, this assay can be utilized to measure renin activity in kidney or other tissues. It is also possible that under low renin concentrations, the incubation time may need to be extended in order to get measurable AngI levels [6]. Since radiolabels require the investigator to have specialized protocols for handling and disposing of radioactive waste, more biomedical companies have pursued other alternatives, such as colorimetric kits. Again, it is up to the user to develop the proper positive and negative controls to determine the specificity and accuracy of these kits.

1.2 ACE

Angiotensin-converting enzyme 1 (ACE1, 150–180 kDa) is a metalloprotease that cleaves two amino acids from AngI to form angiotensin 1-8 (Ang-(1-8) or AngII) [7]. This dipeptidyl carboxypeptidase has a membrane-bound as well as a cleaved form that can be found in the circulation [8–10]. It is a glycosylated protein that can also metabolize other substrates such as bradykinin, substance P, and angiotensin-(1-7) (Ang-(1-7)) [11–14]. ACE can be found in the lung and the endothelium of blood vessels, and a shorter isoform can be found in the testis [15, 16]. ACE binds to two zinc atoms located within HEMGH domains where the histidines help in coordination of the zinc atoms. Typically, ACE activity is measured by using a peptide substrate, such as hippuryl-His-Leu in a solution that contains zinc chloride in a sodium borate buffer at an optimal pH [17]. Since EDTA is a chelating agent, then ACE activity is measured in serum. In using this assay, a set of duplicates are made with incubation of the ACE inhibitor, captopril. The differences between the samples will allow the user to determine the ACE-dependent activity in a sample. It should be noted that mRNA and protein can be measured as well for ACE; however, the activity of ACE is the best indicator for RAAS activation. It should be noted that ACE is also thought to act as a mechano-sensor within the aorta [18]. ACE has been shown to have RGD domains that can bind to integrins and can be mechano-sensors in shear stress or extracellular matrix disruption [19]. Typically, increases in ACE mRNA and protein can be found in aortas that are smooth muscle deficient in fibuilin-4 [18]. These functions could be independent of the catalytic function of ACE and should also be considered when looking at cardiovascular disease.

1.3 ACE2

Angiotensin-converting enzyme 2 (ACE2, 90–120 kDa) is also a metalloprotease, however unlike ACE, ACE2 cleaves only one amino acid from AngII to make Ang-(1-7) [20–22]. ACE2 is also insensitive to ACEi [8]. This monocarboxypeptidase can be cleaved from the membrane to make a soluble form [23]. ACE2 can be found in the kidneys, heart, adipose tissue, blood vessels, brain, and sex organs [21]. ACE2 can be measured using either a radiolabeled AngII or fluorogenic substrate. The ACE2 measurement by using ^{125}I-AngII requires the use of an HPLC so that the radiolabeled Ang-(1-7) fraction can be separated from the other angiotensin peptides. Use of a fluorogenic substrate requires less equipment and money to test for ACE2 activity; however, the difference in substrate affinities must be accounted for as well. In Chapter 6 of this book, our lab uses the fluorogenic approach to measure ACE2 activity in saline or AngII-treated abdominal aortas of mice that are wild type or ACE2-deficient. It must be noted that again in order to determine ACE2 specific activity that either animal knockouts or samples treated with an ACE2 inhibitor (e.g., MLN-4760) must be employed. ACE2 also has the ability to cleave other substrates such as apelin, dynorphin A, and des-Arg9-bradykinin. Again, ACE2 mRNA and protein can be measured; however, the enzymatic activity is the best indicator for activation. ACE2 also has RGD domains that can bind to integrins in cardiomyocytes [19], and overexpression of ACE2 in the heart can protect against ischemia-induced pathophysiology [24].

1.4 Neprilysin

Neprilysin (90–100 kDa) can process AngI directly to Ang-(1-7) [14, 25–28]. Typically, this peptidase goes up with chronic ACE inhibition and can play a significant role in heart failure [29]. Neprilysin requires zinc to function and also has other substrates such as enkephalins, atrial natriuretic peptide, endothelin, substance P, and bradykinin. Neprilysin has been localized to the kidney, heart, and blood vessels, and recently this endopeptidase has been located in the mitochondria of the kidney [30]. In Chapter 9, Dr. Chappell and colleagues discuss important pathways localized to both the mitochondria and the nucleus for the RAS. Neprilysin can be measured using ^{125}I-AngI as a substrate and then measuring ^{125}I-Ang-(1-7) by HPLC. There are also fluorescent substrates for measurement of neprilysin, but neprilysin inhibitors such as thiorphan and phosphoramidon should be used in these measurements to determine neprilysin-specific activity.

1.5 Angiotensinogen

Angiotensinogen is roughly a 450 amino acid protein (55–60 kDa) that when cleaved by renin forms the decapeptide, AngI. It is a glycoprotein that requires species-specific renin to cleave angiotensinogen since mouse renin does not cleave human angiotensinogen. Angiotensinogen can be found primarily in the liver, but has also been found in the kidney, adipose, and other tissues [31–34]. Angiotensinogen release can be stimulated by the acute inflammatory response, insulin, estrogen, glucocorticoids, and thyroid hormone [35–38]. Angiotensinogen can also be elevated in women with pre-eclampsia, and it has been suggested that the oxidized form of angiotensinogen can be readily cleaved by renin in comparison to the reduced form [39]. Since angiotensinogen is a protein and not an enzyme, then Western blot or enzyme-linked immunosorbent assay (ELISA) are preferred methods.

1.6 AT1R, AT2R, and MasR (GPCRs)

The G-protein-coupled receptors (GPCRs), such as AT1R, AT2R, and MasR, have numerous functions in the cell spanning from activating the contractile apparatus in smooth muscle cells, inducing oxidative stress or influencing nitric oxide (NO) production, and influencing cellular proliferation and apoptosis [40–42]. Two independent labs were responsible for the initial cloning of the AT1R [43, 44], and in rodents there are two subtypes of the AT1R, AT1aR and AT1bR [45]. These receptors have similar effects in mice with both receptors playing a role in growth and blood pressure regulation [46–48]. Unfortunately, antibodies targeting the AT1aR have been unsuccessful in development and therefore most labs utilize real-time PCR for the measurement of these receptors [49]. Another area of research is exploring ligands for the AT1R and other angiotensin receptors to modulate downstream signaling. Beta-arrestin has been shown to couple to AT1aR and can modulate angiotensin signaling [50–52]. In Chapter 13 of this book, Bragina and colleagues have developed a fluorescence-based binding assay for screening ligands to the AT1R or other angiotensin receptors. This method allows for a high throughput of compounds that can be screened initially as well as mutational studies that can test for binding affinity of test compounds.

AT2R was first described in fetal and neonatal brain tissue [53]. While expression levels typically decrease after birth, it has been shown that adult mice deficient in AT2R have increased systolic blood pressure and low levels of bradykinin [54]. Recent studies have also shown that agonists for the AT2R, such as Compound-21, can induce natriuresis and lower blood pressure [55]. The AT2R is thought to oppose the function of the AT1R and has been found to reduce atherosclerosis [56], autoimmune disease [57], obesity-hypertension [58], and stroke [59].

The MasR was first originally described as an angiotensin receptor [60]. It was not until 2003 that Ang-(1-7) was identified as the endogenous ligand for the MasR [61]. Since that time, there have been numerous publications showing that the MasR is involved in vasodilation [62], reproduction [63, 64], heart contractility [65], and lipid profiles [66, 67]. While antibodies have been used to target AT2R and MasR via Western blotting, more evidence is needed to prove that these antibodies are specific to these GPCRs.

1.7 Angiotensin Peptides

AngII is considered the central peptide of the RAAS. Since its initial discovery in 1940 [68, 69], numerous other peptides have now been characterized to be a part of this system. In Chapter 14, Olkowicz and colleagues have designed a protocol that helps in the characterization of these peptides using nano-liquid chromatography coupled with mass spectrometry. Peptides, such as Ang-(1-7), Ang-(1-9), angiotensin A, and alamandine, have been shown to be the counter-regulatory arm of the AngII/AT1R pathway and have been manipulated to block the effects of AngII [70–77]. Recent evidence has indicated that angiotensin peptides have different compositions depending on the tissue localization [33, 78–82]. In Chapter 15, Grobe and Elased developed a method for the analysis of angiotensin peptides in the kidney using mass spectrometry. Interestingly, you can see that different peptides are generated in different areas of the kidney indicating specific localization of enzymes necessary for this complex pattern

(Figure 4 of Chapter 15). These advanced techniques can help to provide better information on the angiotensin peptide environment under certain pathophysiological conditions and bring about other peptidase inhibitors that may provide better efficacy against certain types of diseases.

1.8 Aldosterone

The RAAS, as well as increases in extracellular potassium ions, can increase aldosterone secretion. Aldosterone secretion from the adrenals can then act upon the kidneys to conserve sodium and excrete potassium. It also causes increases in water retention and thereby can increase blood pressure. This increase in blood pressure is slower in onset than AngII and the effects it can have on vasoconstriction of blood vessels. If aldosterone secretion is high, such as in primary aldosteronism, renin can then be suppressed in the kidney. While this effect is RAAS-independent, hypertensive patients typically exhibit an over-active RAAS. In Chapter 12, Liu et al., have shown that aldosterone in the presence of salt can increase blood pressure and cause abdominal aortic aneurysms in aged C57BL/6J mice [83]. This new model gives investigators new opportunities to study cardiovascular diseases independent of AngII. Aldosterone is a steroid hormone that is derived from cholesterol. Most assays use either mass-spectrometry or ELISA-based methods for detection [84, 85]. Aldosterone binds to the mineralocorticoid receptor (MR) and this receptor has been localized to tissues such as kidney, liver, brain, ileum, brown and white adipose tissue, heart, and blood vessels [83, 86–93].

1.9 RAAS and Energy Balance

Obesity and metabolic disorders can be a result of an imbalance of energy intake versus energy expenditure. Recent work has shown that leptin and the AT1aR interact in the regions of the brain, such as the subfornical organ (SFO) [94, 95]. This reaction can induce an increase in sympathetic nerve activity (SNA) which can result in increased thermogenesis in brown adipose tissue and browning of white adipose tissue (beige phenotype) [96, 97]. The creation of a double transgenic mouse, referred to as the sRA mouse, expresses human renin controlled by the neuron-specific synapsin promoter and the human angiotensinogen gene controlled by its own promoter. Since there is species-specific cleavage of angiotensinogen by renin, this allows for increases in brain RAAS activity that is not influenced by the peripheral RAAS [96, 98]. Recent data has shown that the sRA mouse has elevated energy expenditure and thermogenesis [96]. This effect has been shown to be mediated by elevated sympathetic activity; however, there was no reported increase in uncoupling protein-1 (UCP-1) [96]. These changes in metabolic regulation and increased sympathetic drive could help to explain some facets of obesity-hypertension and in Chapter 10, Grobe describes in great detail the proper techniques needed to examine energy balance in mice.

1.10 RAAS and Oxidative Stress

Early studies were first able to show that AngII could induce superoxide production in hypertensive rats and that extracellular superoxide dismutase (SOD) mRNA production could be stabilized by AngII [99, 100]. Studies have also shown that AngII can activate hydrogen peroxide (H_2O_2) pathways, such as p38 mitogen-activated protein kinase (p38MAPK) that can increase vascular hypertrophy [101]. These studies have been translated to human vascular smooth muscle cells as well and indicate that oxidative stress can be mediated through AngII and that inhibitors of phospholipase D and NADPH oxidase can block these effects [102]. Oxidative stress has also been linked to AngII-induced inflammation [103]. Specifically, it has been shown that AngII can induce JAK-STAT phosphorylation which results in increases in p47 phox, a subunit of NADPH oxidase and interleukin-6 (IL-6) transcription [103]. These effects could be blocked with the use of an NADPH oxidase inhibitor (e.g., diphenyleneiodonium, DPI) [103]. Whether through JAK-STAT signaling or other types of tyrosine kinases, AngII can stimulate these pathways through redox-sensitive mechanisms [104–106]. Likewise, ACE2 has been shown to inhibit oxidative stress through both JAK-STAT and MAPK pathways [107]. In Chapter 3, Tsiropoulou and Touyz have developed a method to measure protein carbonylation and protein tyrosine phosphatase oxidation in vascular smooth muscle cells. This technique should help to discern differences between signaling of these redox-sensitive pathways in AngII-related diseases.

1.11 RAAS and Atherosclerosis

The use of ACE inhibitors has been shown to reduce atherosclerosis in a number of animal models [108–110]. Likewise, it has been shown that AngII signaling goes up with hypercholesterolemia in rabbit and rodent models [111–113]. Newly developed drugs, such as aliskiren, have also shown decreases in atherosclerosis [114, 115]. While these drugs have shown benefit in humans, the use of combination therapy, through the use of an ARB and ACEi, has not shown any additional benefit in cardiovascular outcomes [116]. Recently, the ACE2/Ang-(1-7)/MasR pathway has also shown a beneficial role in decreasing atherosclerosis in animal models [73, 117–119]. While Ang-(1-7) co-infusion with AngII or through diet alone has shown decreases in atherosclerosis [73, 120], the MasR has been shown to either be detrimental [121] or no change [67] in the development of atherosclerosis. Also, downstream components in this arm, such as MrgD and alamandine, are also found in the aorta and involved in vessel reactivity, respectively [122], however it still has not been determined the role, if any, these components might play in atherosclerosis progression or regression. Recent evidence does suggest that the RAAS can also be detected in human atherosclerosis [123–126]. Methodology to measure atherosclerosis can be performed in the aortic arch by en face analysis or in the aortic root [127]. Quantification can be performed in the aortic root by lesion area; however, this does not give any information on lesion composition. In Chapter 2 of this book, Wu and colleagues use a color segmentation method to quantify smooth muscle and macrophage accumulation within specific lesions. This method should also add benefit to determining plaque stability as well in these mouse models of atherosclerosis.

1.12 RAAS and Stem Cells

It has been shown that a hematopoietic peptide called N-acetyl-seryl-aspartyl-lysyl-proline (Ac-SDKP) could be hydrolyzed in vitro by ACE1 and that this peptide can block stem cell proliferation [128, 129]. It was determined that the N-terminal active site of ACE1 could hydrolyze Ac-SDKP 50 times faster than the C-terminal active site [129]. This led to the idea of a local RAAS in the bone marrow [130]. In 1997, Mrug et al., showed that AngII could stimulate proliferation of CD34+ hematopoietic progenitor cells and that this effect could be blocked by losartan [131]. Since this time, other components such as renin [132], ACE2 [133–135], angiotensinogen [136], MasR/Ang-(1-7) [137], and AT1aR [138] have also been found in bone marrow. Stem cells have gained tremendous attention as possible therapies for a number of different diseases. In Chapter 4 of this book, new methodology for studying the RAAS in stem cells and how to properly gate the cells using flow cytometry are described by Dr. Yagna Jarajapu.

1.13 RAAS and Hypertension

The importance of the kidney and regulation of blood pressure was first described by Harry Goldblatt [139, 140]. These experiments were typically conducted in dogs, monkeys, and other larger mammals, however with the development of gene knockout models in rodents, particularly mice; it became apparent that technology was needed for blood pressure measurements in smaller mammals. One of the first methods described for measuring blood pressure in mice was by Hicks and colleagues in Melbourne, Australia [141]. Since this time, procedures such as tail-cuff and radiotelemetry have been used to measure blood pressure in mice. In Chapters 6 and 7 of this book, Yu et al., have outlined the basic procedures for both using tail-cuff platforms and radiotelemetry for mice. These methods allow users to best discern which methodology would work best for his/her given experiments. Radiotelemetry is considered the "gold standard" when measuring blood pressure in mice and allows the user to get more information on activity and circadian rhythms of systolic and diastolic blood pressure compared to tail-cuff measurements [142]. A number of genetic knockout mouse models have shown that the RAAS is involved in the maintenance of blood pressure [31, 47, 143–146]. Stimuli, such as high salt, saturated fat diets, and infusion of AngII, can be used to increase blood pressure and activate the RAAS. When utilizing these methods, it is important to acclimate the mice before recording the blood pressure, since stress and glucocorticoid release can also influence the blood pressure response [147, 148]. Other considerations that can influence the blood pressure are age, mouse strain, and sex. In fact, numerous studies have shown sex-dependent effects of blood pressure with RAAS activation [145, 149–155]. While these effects can be hormone driven, new evidence now indicates that sex chromosomes can also influence AngII-induced hypertension [156, 157]. It is important for the investigator to understand all of these factors that might influence blood pressure so that he/she can best set up the experiment.

1.14 RAAS and Inflammation

AngII has also been found to be a pro-inflammatory peptide. Early studies have shown that AngII can bind to immune cells in both human and animal models and induce oxidative stress, NADPH oxidase activity, release of cytokines, and increased expression of adhesion markers [158–163]. AngII has also been shown to promote lymphocyte responses that can promote kidney injury [164]. Lymphocytes can be divided into either natural killer (NK) cells, T-cells, or B-cells. T-cells, such as CD4+ or CD8+ lymphocytes, are a part of the adaptive immune system, and can play a role in hypertension, obesity, and diabetes [165–167]. Flow cytometry is a powerful technique that can allow the separation of these immune cells that can be further used in cell culture or for other molecular techniques, such as real-time PCR or Western blotting. In Chapter 8 of this book, Rudemiller and Crowley detail a protocol for the isolation of T-cells in the kidney and the appropriate markers for classes of T-cells along with cytokine products, such as tumor necrosis factor-alpha (TNF-α) and interferon gamma (INF-γ). Recently, it has been shown that Cluster of Differentiation 70 (CD70), a protein expressed in antigen-presenting cells, and CD27 on T-cells play a critical role in the formation of T-memory cells [168, 169]. These T-memory cells can then play an important role in kidney injury and recurrent hypertension [168]. T-cells have also been shown to play a role in RAAS-activated hypertension [170, 171], obesity-associated diabetes [172], and autoimmune disease [173]. The RAAS-activated immune system provides many opportunities to further understand the role of these cells within different types of human diseases.

This is but a brief introduction to the RAAS and different techniques utilized to measure this complex system. It is important to realize the limitations to any research technique and the proper controls needed to insure accurate and reliable results. I hope that the reader finds these chapters very useful in their future studies.

References

1. Tigerstedt R, Bergman PQ (1898) Niere und Kreislaufl. Skand Arch Physiol 8(1):223–271. doi:10.1111/j.1748-1716.1898.tb00272.x
2. Stokes GS, Oates HF, Weber MA (1975) Angiotensin blockade in studies of the feedback control of renin release in rats and rabbits. Clin Sci Mol Med Suppl 2:33s–36s
3. Zanchetti A, Stella A, Leonetti G, Morganti A, Terzoli L (1976) Control of renin release: a review of experimental evidence and clinical implications. Am J Cardiol 37(4):675–691
4. Chappell MC (2016) Biochemical evaluation of the renin-angiotensin system: the good, bad, and absolute? Am J Physiol Heart Circ Physiol 310(2):H137–H152. doi:10.1152/ajpheart.00618.2015
5. McDonald KM, Taher S, Aisenbrey G, De Torrente A, Schrier RW (1975) Effect of angiotensin II and an angiotensin II inhibitor on renin secretion in the dog. Am J Phys 228(5):1562–1567
6. Hartman D, Sagnella GA, Chesters CA, Macgregor GA (2004) Direct renin assay and plasma renin activity assay compared. Clin Chem 50(11):2159–2161. doi:10.1373/clinchem.2004.033654
7. Skeggs LT Jr, Kahn JR, Shumway NP (1956) The preparation and function of the hypertensin-converting enzyme. J Exp Med 103(3):295–299
8. Allinson TM, Parkin ET, Condon TP, Schwager SL, Sturrock ED, Turner AJ, Hooper NM (2004) The role of ADAM10 and ADAM17 in the ectodomain shedding of angiotensin converting enzyme and the amyloid precursor protein. Eur J Biochem 271(12):2539–2547. doi:10.1111/j.1432-1033.2004.04184.x
9. Woodman ZL, Oppong SY, Cook S, Hooper NM, Schwager SL, Brandt WF, Ehlers MR, Sturrock ED (2000) Shedding of somatic angiotensin-converting enzyme (ACE) is inefficient compared with testis ACE despite cleavage at identical stalk sites. Biochem J 347(Pt 3):711–718
10. Danilov SM, Gordon K, Nesterovitch AB, Lunsdorf H, Chen Z, Castellon M, Popova IA, Kalinin S, Mendonca E, Petukhov PA, Schwartz DE, Minshall RD, Sturrock ED (2011) An angiotensin I-converting enzyme mutation (Y465D) causes a dramatic increase in blood ACE via accelerated ACE shedding. PLoS One 6(10):e25952. doi:10.1371/journal.pone.0025952
11. Skidgel RA, Engelbrecht S, Johnson AR, Erdos EG (1984) Hydrolysis of substance p and neurotensin by converting enzyme and neutral endopeptidase. Peptides 5(4):769–776
12. Kohara K, Brosnihan KB, Ferrario CM (1993) Angiotensin(1-7) in the spontaneously hypertensive rat. Peptides 14(5):883–891
13. Campbell WG Jr, Donohue JA, Duket LH (1980) Alterations in responses to bradykinin, angiotensin I, and angiotensin II during the induction phase of one-kidney, one-wrapped hypertension and associated arterial disease in rabbits. Am J Pathol 98(2):457–484
14. Rice GI, Thomas DA, Grant PJ, Turner AJ, Hooper NM (2004) Evaluation of angiotensin-converting enzyme (ACE), its homologue ACE2 and neprilysin in angiotensin peptide metabolism. Biochem J 383(Pt 1):45–51. doi:10.1042/BJ20040634
15. Bernstein KE, Martin BM, Bernstein EA, Linton J, Striker L, Striker G (1988) The isolation of angiotensin-converting enzyme cDNA. J Biol Chem 263(23):11021–11024
16. Howard TE, Shai SY, Langford KG, Martin BM, Bernstein KE (1990) Transcription of testicular angiotensin-converting enzyme (ACE) is initiated within the 12th intron of the somatic ACE gene. Mol Cell Biol 10(8):4294–4302
17. Brecher P, Tercyak A, Chobanian AV (1981) Properties of angiotensin-converting enzyme in intact cerebral microvessels. Hypertension 3(2):198–204
18. Yamashiro Y, Papke CL, Kim J, Ringuette LJ, Zhang QJ, Liu ZP, Mirzaei H, Wagenseil JE, Davis EC, Yanagisawa H (2015) Abnormal mechanosensing and cofilin activation promote the progression of ascending aortic aneurysms in mice. Sci Signal 8(399):ra105. doi:10.1126/scisignal.aab3141
19. Clarke NE, Fisher MJ, Porter KE, Lambert DW, Turner AJ (2012) Angiotensin converting enzyme (ACE) and ACE2 bind integrins and ACE2 regulates integrin signalling. PLoS One 7(4):e34747. doi:10.1371/journal.pone.0034747. PONE-D-11-26032 [pii]
20. Tipnis SR, Hooper NM, Hyde R, Karran E, Christie G, Turner AJ (2000) A human homolog of angiotensin-converting enzyme. Cloning and functional expression as a captopril-insensitive carboxypeptidase. J Biol Chem 275(43):33238–33243. doi:10.1074/jbc.M002615200. M002615200 [pii]
21. Donoghue M, Hsieh F, Baronas E, Godbout K, Gosselin M, Stagliano N, Donovan M, Woolf B, Robison K, Jeyaseelan R, Breitbart RE, Acton S (2000) A novel angiotensin-converting

enzyme-related carboxypeptidase (ACE2) converts angiotensin I to angiotensin 1-9. Circ Res 87(5):E1–E9

22. Vickers C, Hales P, Kaushik V, Dick L, Gavin J, Tang J, Godbout K, Parsons T, Baronas E, Hsieh F, Acton S, Patane M, Nichols A, Tummino P (2002) Hydrolysis of biological peptides by human angiotensin-converting enzyme-related carboxypeptidase. J Biol Chem 277(17):14838–14843. doi:10.1074/jbc.M200581200. M200581200 [pii]
23. Lambert DW, Yarski M, Warner FJ, Thornhill P, Parkin ET, Smith AI, Hooper NM, Turner AJ (2005) Tumor necrosis factor-alpha convertase (ADAM17) mediates regulated ectodomain shedding of the severe-acute respiratory syndrome-coronavirus (SARS-CoV) receptor, angiotensin-converting enzyme-2 (ACE2). J Biol Chem 280(34):30113–30119. doi:10.1074/jbc.M505111200
24. Der Sarkissian S, Grobe JL, Yuan L, Narielwala DR, Walter GA, Katovich MJ, Raizada MK (2008) Cardiac overexpression of angiotensin converting enzyme 2 protects the heart from ischemia-induced pathophysiology. Hypertension 51(3):712–718. doi:10.1161/HYPERTENSIONAHA.107.100693
25. Welches WR, Brosnihan KB, Ferrario CM (1993) A comparison of the properties and enzymatic activities of three angiotensin processing enzymes: angiotensin converting enzyme, prolyl endopeptidase and neutral endopeptidase 24.11. Life Sci 52(18):1461–1480
26. Iyer SN, Ferrario CM, Chappell MC (1998) Angiotensin-(1-7) contributes to the antihypertensive effects of blockade of the renin-angiotensin system. Hypertension 31(1 Pt 2):356–361
27. Ferrario CM, Chappell MC, Dean RH, Iyer SN (1998) Novel angiotensin peptides regulate blood pressure, endothelial function, and natriuresis. J Am Soc Nephrol 9(9):1716–1722
28. Ishida M, Ogawa M, Kosaki G, Mega T, Ikenaka T (1983) Purification and characterization of the neutral endopeptidase from human kidney. J Biochem 94(1):17–24
29. Lim GB (2014) Heart failure: LCZ696—a PARADIGM shift in treatment for heart failure. Nat Rev Cardiol 11(11):618. doi:10.1038/nrcardio.2014.139
30. Wilson BA, Nautiyal M, Gwathmey TM, Rose JC, Chappell MC (2015) Evidence for a mitochondrial angiotensin-(1-7) system in the kidney. Am J Physiol Renal Physiol 00479:02015. doi:10.1152/ajprenal.00479.2015
31. Yiannikouris F, Wang Y, Shoemaker R, Larian N, Thompson J, English VL, Charnigo R, Su W, Gong M, Cassis LA (2015) Deficiency of angiotensinogen in hepatocytes markedly decreases blood pressure in lean and obese male mice. Hypertension 66(4):836–842. doi:10.1161/HYPERTENSIONAHA.115.06040
32. Yiannikouris F, Gupte M, Putnam K, Thatcher S, Charnigo R, Rateri DL, Daugherty A, Cassis LA (2012) Adipocyte deficiency of angiotensinogen prevents obesity-induced hypertension in male mice. Hypertension 60(6):1524–1530. doi:10.1161/HYPERTENSIONAHA.112.192690
33. Dzau VJ, Brody T, Ellison KE, Pratt RE, Ingelfinger JR (1987) Tissue-specific regulation of renin expression in the mouse. Hypertension 9(6 Pt 2):III36–III41
34. Gomez RA, Cassis L, Lynch KR, Chevalier RL, Wilfong N, Carey RM, Peach MJ (1988) Fetal expression of the angiotensinogen gene. Endocrinology 123(5):2298–2302. doi:10.1210/endo-123-5-2298
35. Itoh N, Matsuda T, Ohtani R, Okamoto H (1989) Angiotensinogen production by rat hepatoma cells is stimulated by B cell stimulatory factor 2/interleukin-6. FEBS Lett 244(1):6–10
36. Murakami E, Hiwada K, Kokubu T (1980) Effects of insulin and glucagon on production of renin substrate by the isolated rat liver. J Endocrinol 85(1):151–153
37. Krakoff LR, Eisenfeld AJ (1977) Hormonal control of plasma renin substrate; (angiotensinogen). Circ Res 41(4 Suppl 2):43–46
38. Dzau VJ, Herrmann HC (1982) Hormonal control of angiotensinogen production. Life Sci 30(7–8):577–584
39. Zhou A, Carrell RW, Murphy MP, Wei Z, Yan Y, Stanley PL, Stein PE, Broughton Pipkin F, Read RJ (2010) A redox switch in angiotensinogen modulates angiotensin release. Nature 468(7320):108–111. doi:10.1038/nature09505
40. Ryan MJ, Didion SP, Mathur S, Faraci FM, Sigmund CD (2004) Angiotensin II-induced vascular dysfunction is mediated by the AT1A receptor in mice. Hypertension 43(5):1074–1079. doi:10.1161/01.HYP.0000123074.89717.3d
41. Mehta PK, Griendling KK (2007) Angiotensin II cell signaling: physiological and pathological effects in the cardiovascular system. Am J Physiol Cell Physiol 292(1):C82–C97. doi:10.1152/ajpcell.00287.2006
42. Simoes e Silva AC, Silveira KD, Ferreira AJ, Teixeira MM (2013) ACE2, angiotensin-(1-7) and mas receptor axis in inflammation and fibrosis. Br J Pharmacol 169(3):477–492. doi:10.1111/bph.12159

43. Murphy TJ, Alexander RW, Griendling KK, Runge MS, Bernstein KE (1991) Isolation of a cDNA encoding the vascular type-1 angiotensin II receptor. Nature 351(6323):233–236. doi:10.1038/351233a0
44. Sasaki K, Yamano Y, Bardhan S, Iwai N, Murray JJ, Hasegawa M, Matsuda Y, Inagami T (1991) Cloning and expression of a complementary DNA encoding a bovine adrenal angiotensin II type-1 receptor. Nature 351(6323):230–233. doi:10.1038/351230a0
45. Burson JM, Aguilera G, Gross KW, Sigmund CD (1994) Differential expression of angiotensin receptor 1A and 1B in mouse. Am J Phys 267(2 Pt 1):E260–E267
46. Ito M, Oliverio MI, Mannon PJ, Best CF, Maeda N, Smithies O, Coffman TM (1995) Regulation of blood pressure by the type 1A angiotensin II receptor gene. Proc Natl Acad Sci U S A 92(8):3521–3525
47. Oliverio MI, Best CF, Kim HS, Arendshorst WJ, Smithies O, Coffman TM (1997) Angiotensin II responses in AT1A receptor-deficient mice: a role for AT1B receptors in blood pressure regulation. Am J Phys 272(4 Pt 2):F515–F520
48. Poduri A, Owens AP 3rd, Howatt DA, Moorleghen JJ, Balakrishnan A, Cassis LA, Daugherty A (2012) Regional variation in aortic AT1b receptor mRNA abundance is associated with contractility but unrelated to atherosclerosis and aortic aneurysms. PLoS One 7(10):e48462. doi:10.1371/journal.pone.0048462
49. Herrera M, Sparks MA, Alfonso-Pecchio AR, Harrison-Bernard LM, Coffman TM (2013) Lack of specificity of commercial antibodies leads to misidentification of angiotensin type 1 receptor protein. Hypertension 61(1):253–258. doi:10.1161/HYPERTENSIONAHA.112.203679
50. Zhang J, Barak LS, Anborgh PH, Laporte SA, Caron MG, Ferguson SS (1999) Cellular trafficking of G protein-coupled receptor/beta-arrestin endocytic complexes. J Biol Chem 274(16):10999–11006
51. Rajagopal K, Whalen EJ, Violin JD, Stiber JA, Rosenberg PB, Premont RT, Coffman TM, Rockman HA, Lefkowitz RJ (2006) Beta-arrestin2-mediated inotropic effects of the angiotensin II type 1A receptor in isolated cardiac myocytes. Proc Natl Acad Sci U S A 103(44):16284–16289. doi:10.1073/pnas.0607583103
52. Rakesh K, Yoo B, Kim IM, Salazar N, Kim KS, Rockman HA (2010) Beta-arrestin-biased agonism of the angiotensin receptor induced by mechanical stress. Sci Signal 3(125):ra46. doi:10.1126/scisignal.2000769
53. Nuyt AM, Lenkei Z, Palkovits M, Corvol P, Llorens-Cortes C (1999) Ontogeny of angiotensin II type 2 receptor mRNA expression in fetal and neonatal rat brain. J Comp Neurol 407(2):193–206
54. Siragy HM, Inagami T, Ichiki T, Carey RM (1999) Sustained hypersensitivity to angiotensin II and its mechanism in mice lacking the subtype-2 (AT2) angiotensin receptor. Proc Natl Acad Sci U S A 96(11):6506–6510
55. Kemp BA, Howell NL, Gildea JJ, Keller SR, Padia SH, Carey RM (2014) AT(2) receptor activation induces natriuresis and lowers blood pressure. Circ Res 115(3):388–399. doi:10.1161/CIRCRESAHA.115.304110
56. Sales VL, Sukhova GK, Lopez-Ilasaca MA, Libby P, Dzau VJ, Pratt RE (2005) Angiotensin type 2 receptor is expressed in murine atherosclerotic lesions and modulates lesion evolution. Circulation 112(21):3328–3336. doi:10.1161/CIRCULATIONAHA.105.541714
57. Valero-Esquitino V, Lucht K, Namsolleck P, Monnet-Tschudi F, Stubbe T, Lucht F, Liu M, Ebner F, Brandt C, Danyel LA, Villela DC, Paulis L, Thoene-Reineke C, Dahlof B, Hallberg A, Unger T, Sumners C, Steckelings UM (2015) Direct angiotensin type 2 receptor (AT2R) stimulation attenuates T-cell and microglia activation and prevents demyelination in experimental autoimmune encephalomyelitis in mice. Clin Sci (Lond) 128(2):95–109. doi:10.1042/CS20130601
58. Ali Q, Wu Y, Hussain T (2013) Chronic AT2 receptor activation increases renal ACE2 activity, attenuates AT1 receptor function and blood pressure in obese Zucker rats. Kidney Int 84(5):931–939. doi:10.1038/ki.2013.193
59. Joseph JP, Mecca AP, Regenhardt RW, Bennion DM, Rodriguez V, Desland F, Patel NA, Pioquinto DJ, Unger T, Katovich MJ, Steckelings UM, Sumners C (2014) The angiotensin type 2 receptor agonist compound 21 elicits cerebroprotection in endothelin-1 induced ischemic stroke. Neuropharmacology 81:134–141. doi:10.1016/j.neuropharm.2014.01.044
60. Jackson TR, Blair LA, Marshall J, Goedert M, Hanley MR (1988) The mas oncogene encodes an angiotensin receptor. Nature 335(6189):437–440. doi:10.1038/335437a0
61. Santos RA, Simoes e Silva AC, Maric C, Silva DM, Machado RP, de Buhr I, Heringer-Walther S, Pinheiro SV, Lopes MT, Bader M, Mendes EP, Lemos VS, Campagnole-Santos MJ, Schultheiss HP, Speth R, Walther T (2003) Angiotensin-(1-7) is an endogenous ligand for the G protein-coupled receptor

mas. Proc Natl Acad Sci U S A 100(14):8258–8263. doi:10.1073/pnas.1432869100
62. Lemos VS, Silva DM, Walther T, Alenina N, Bader M, Santos RA (2005) The endothelium-dependent vasodilator effect of the nonpeptide Ang(1-7) mimic AVE 0991 is abolished in the aorta of mas-knockout mice. J Cardiovasc Pharmacol 46(3):274–279
63. Reis AB, Araujo FC, Pereira VM, Dos Reis AM, Santos RA, Reis FM (2010) Angiotensin (1-7) and its receptor mas are expressed in the human testis: implications for male infertility. J Mol Histol 41(1):75–80. doi:10.1007/s10735-010-9264-8
64. Leal MC, Pinheiro SV, Ferreira AJ, Santos RA, Bordoni LS, Alenina N, Bader M, Franca LR (2009) The role of angiotensin-(1-7) receptor mas in spermatogenesis in mice and rats. J Anat 214(5):736–743. doi:10.1111/j.1469-7580.2009.01058.x
65. Santos RA, Castro CH, Gava E, Pinheiro SV, Almeida AP, Paula RD, Cruz JS, Ramos AS, Rosa KT, Irigoyen MC, Bader M, Alenina N, Kitten GT, Ferreira AJ (2006) Impairment of in vitro and in vivo heart function in angiotensin-(1-7) receptor MAS knockout mice. Hypertension 47(5):996–1002. doi:10.1161/01.HYP.0000215289.51180.5c
66. Santos SH, Fernandes LR, Mario EG, Ferreira AV, Porto LC, Alvarez-Leite JI, Botion LM, Bader M, Alenina N, Santos RA (2008) Mas deficiency in FVB/N mice produces marked changes in lipid and glycemic metabolism. Diabetes 57(2):340–347. doi:10.2337/db07-0953
67. Silva AR, Aguilar EC, Alvarez-Leite JI, da Silva RF, Arantes RM, Bader M, Alenina N, Pelli G, Lenglet S, Galan K, Montecucco F, Mach F, Santos SH, Santos RA (2013) Mas receptor deficiency is associated with worsening of lipid profile and severe hepatic steatosis in ApoE-knockout mice. Am J Physiol Regul Integr Comp Physiol 305(11):R1323–R1330. doi:10.1152/ajpregu.00249.2013
68. Page IH, Helmer OM (1940) A crystalline pressor substance (Angiotonin) resulting from the reaction between renin and renin-activator. J Exp Med 71(1):29–42
69. Menendez EB, Fasciolo JC, Houssay BA, Leloir LF, Munoz JM, Taquini AC (1943) Angiotonin or hypertensin. Science 98(2553):495. doi:10.1126/science.98.2553.495
70. Zhang F, Ren X, Zhao M, Zhou B, Han Y (2016) Angiotensin-(1-7) abrogates angiotensin II-induced proliferation, migration and inflammation in VSMCs through inactivation of ROS-mediated PI3K/Akt and MAPK/ERK signaling pathways. Sci Rep 6:34621. doi:10.1038/srep34621
71. Sampaio WO, Henrique de Castro C, Santos RA, Schiffrin EL, Touyz RM (2007) Angiotensin-(1-7) counterregulates angiotensin II signaling in human endothelial cells. Hypertension 50(6):1093–1098. doi:10.1161/HYPERTENSIONAHA.106.084848 HYPERTENSIONAHA.106.084848 [pii]
72. Hayashi N, Yamamoto K, Ohishi M, Tatara Y, Takeya Y, Shiota A, Oguro R, Iwamoto Y, Takeda M, Rakugi H (2010) The counter-regulating role of ACE2 and ACE2-mediated angiotensin 1-7 signaling against angiotensin II stimulation in vascular cells. Hypertens Res 33(11):1182–1185. doi:10.1038/hr.2010.147. hr2010147 [pii]
73. Thatcher SE, Zhang X, Howatt DA, Lu H, Gurley SB, Daugherty A, Cassis LA (2011) Angiotensin-converting enzyme 2 deficiency in whole body or bone marrow-derived cells increases atherosclerosis in low-density lipoprotein receptor-/- mice. Arterioscler Thromb Vasc Biol 31(4):758–765. doi:10.1161/atvbaha.110.221614
74. Zhang F, Hu Y, Xu Q, Ye S (2010) Different effects of angiotensin II and angiotensin-(1-7) on vascular smooth muscle cell proliferation and migration. PLoS One 5(8):e12323. doi:10.1371/journal.pone.0012323
75. Lautner RQ, Villela DC, Fraga-Silva RA, Silva N, Verano-Braga T, Costa-Fraga F, Jankowski J, Jankowski V, Sousa F, Alzamora A, Soares E, Barbosa C, Kjeldsen F, Oliveira A, Braga J, Savergnini S, Maia G, Peluso AB, Passos-Silva D, Ferreira A, Alves F, Martins A, Raizada M, Paula R, Motta-Santos D, Klempin F, Pimenta A, Alenina N, Sinisterra R, Bader M, Campagnole-Santos MJ, Santos RA (2013) Discovery and characterization of alamandine: a novel component of the renin-angiotensin system. Circ Res 112(8):1104–1111. doi:10.1161/CIRCRESAHA.113.301077
76. Chen Z, Tan F, Erdos EG, Deddish PA (2005) Hydrolysis of angiotensin peptides by human angiotensin I-converting enzyme and the resensitization of B2 kinin receptors. Hypertension 46(6):1368–1373. doi:10.1161/01.HYP.0000188905.20884.63
77. Coutinho DC, Foureaux G, Rodrigues KD, Salles RL, Moraes PL, Murca TM, De Maria ML, Gomes ER, Santos RA, Guatimosim S, Ferreira AJ (2014) Cardiovascular effects of angiotensin A: a novel peptide of the renin-angiotensin system. J Renin-Angiotensin-Aldosterone Syst 15(4):480–486. doi:10.1177/1470320312474856
78. Grobe N, Weir NM, Leiva O, Ong FS, Bernstein KE, Schmaier AH, Morris M, Elased KM (2013) Identification of prolyl carboxypeptidase as an alternative enzyme

for processing of renal angiotensin II using mass spectrometry. Am J Physiol Cell Physiol 304(10):C945–C953. doi:10.1152/ajpcell.00346.2012

79. Grobe N, Elased KM, Cool DR, Morris M (2012) Mass spectrometry for the molecular imaging of angiotensin metabolism in kidney. Am J Physiol Endocrinol Metab 302(8):E1016–E1024. doi:10.1152/ajpendo.00515.2011
80. Carey RM, Levens NR, Peach MJ (1980) Studies of the functional role of the intrarenal renin-angiotensin system. Prog Biochem Pharmacol 17:6–13
81. Cassis LA, Saye J, Peach MJ (1988) Location and regulation of rat angiotensinogen messenger RNA. Hypertension 11(6 Pt 2):591–596
82. Schelling P, Ganten U, Sponer G, Unger T, Ganten D (1980) Components of the renin-angiotensin system in the cerebrospinal fluid of rats and dogs with special consideration of the origin and the fate of angiotensin II. Neuroendocrinology 31(5):297–308
83. Liu S, Xie Z, Daugherty A, Cassis LA, Pearson KJ, Gong MC, Guo Z (2013) Mineralocorticoid receptor agonists induce mouse aortic aneurysm formation and rupture in the presence of high salt. Arterioscler Thromb Vasc Biol 33(7):1568–1579. doi:10.1161/ATVBAHA.112.300820
84. Schwartz F, Hadas E, Harnik M, Solomon B (1990) Enzyme-linked immunosorbent assays for determination of plasma aldosterone using highly specific polyclonal antibodies. J Immunoass 11(2):215–234. doi:10.1080/01971529008053270
85. Prome D, Viger A, Marquet A (1988) Use of tandem mass spectrometry (MS-MS) for aldosterone assay at the nanogram level in complex biological mixtures. Anal Biochem 172(1):264–269
86. Lombes M, Farman N, Oblin ME, Baulieu EE, Bonvalet JP, Erlanger BF, Gasc JM (1990) Immunohistochemical localization of renal mineralocorticoid receptor by using an anti-idiotypic antibody that is an internal image of aldosterone. Proc Natl Acad Sci U S A 87(3):1086–1088
87. Van Eekelen JA, Jiang W, De Kloet ER, Bohn MC (1988) Distribution of the mineralocorticoid and the glucocorticoid receptor mRNAs in the rat hippocampus. J Neurosci Res 21(1):88–94. doi:10.1002/jnr.490210113
88. Fukushima K, Sasano H, Sasaki I, Nagura H, Funayama Y, Matsuno S (1994) Increased expression of mineralocorticoid receptor in human ileum after total colectomy: immunohistochemical and immunoblotting studies. Tohoku J Exp Med 173(4):383–390
89. Zennaro MC, Le Menuet D, Viengchareun S, Walker F, Ricquier D, Lombes M (1998) Hibernoma development in transgenic mice identifies brown adipose tissue as a novel target of aldosterone action. J Clin Invest 101(6):1254–1260. doi:10.1172/JCI1915
90. Liu G, Grifman M, Keily B, Chatterton JE, Staal FW, Li QX (2006) Mineralocorticoid receptor is involved in the regulation of genes responsible for hepatic glucose production. Biochem Biophys Res Commun 342(4):1291–1296. doi:10.1016/j.bbrc.2006.02.065
91. Carruthers BM, Ledray RD, Seraglia M, McIntosh HW, Walsh GC (1963) Effect of an aldosterone antagonist (spironolactone) on patients with severe congestive heart failure. Can Med Assoc J 89:633–641
92. Lazar G, Pagano M, Agarwal MK (1990) Purification and characterization of the activated mineralocorticoid receptor from rat myocardium. Biochim Biophys Acta 1033(1):41–48
93. Kornel L, Kanamarlapudi N, Travers T, Taff DJ, Patel N, Chen C, Baum RM, Raynor WJ (1982) Studies on high affinity binding of mineralo- and glucocorticoids in rabbit aorta cytosol. J Steroid Biochem 16(2):245–264
94. Hilzendeger AM, Morgan DA, Brooks L, Dellsperger D, Liu X, Grobe JL, Rahmouni K, Sigmund CD, Mark AL (2012) A brain leptin-renin angiotensin system interaction in the regulation of sympathetic nerve activity. Am J Physiol Heart Circ Physiol 303(2):H197–H206. doi:10.1152/ajpheart.00974.2011
95. Young CN, Morgan DA, Butler SD, Rahmouni K, Gurley SB, Coffman TM, Mark AL, Davisson RL (2015) Angiotensin type 1a receptors in the forebrain subfornical organ facilitate leptin-induced weight loss through brown adipose tissue thermogenesis. Mol Metab 4(4):337–343. doi:10.1016/j.molmet.2015.01.007
96. Grobe JL, Grobe CL, Beltz TG, Westphal SG, Morgan DA, Xu D, de Lange WJ, Li H, Sakai K, Thedens DR, Cassis LA, Rahmouni K, Mark AL, Johnson AK, Sigmund CD (2010) The brain renin-angiotensin system controls divergent efferent mechanisms to regulate fluid and energy balance. Cell Metab 12(5):431–442. doi:10.1016/j.cmet.2010.09.011
97. Contreras C, Nogueiras R, Dieguez C, Medina-Gomez G, Lopez M (2016) Hypothalamus and thermogenesis: heating the BAT, browning the WAT. Mol Cell Endocrinol. doi:10.1016/j.mce.2016.08.002
98. Littlejohn NK, Keen HL, Weidemann BJ, Claflin KE, Tobin KV, Markan KR, Park S,

Naber MC, Gourronc FA, Pearson NA, Liu X, Morgan DA, Klingelhutz AJ, Potthoff MJ, Rahmouni K, Sigmund CD, Grobe JL (2016) Suppression of resting metabolism by the angiotensin AT2 receptor. Cell Rep 16(6):1548–1560. doi:10.1016/j.celrep.2016.07.003

99. Laursen JB, Rajagopalan S, Galis Z, Tarpey M, Freeman BA, Harrison DG (1997) Role of superoxide in angiotensin II-induced but not catecholamine-induced hypertension. Circulation 95(3):588–593
100. Fukai T, Siegfried MR, Ushio-Fukai M, Griendling KK, Harrison DG (1999) Modulation of extracellular superoxide dismutase expression by angiotensin II and hypertension. Circ Res 85(1):23–28
101. Ushio-Fukai M, Alexander RW, Akers M, Griendling KK (1998) p38 mitogen-activated protein kinase is a critical component of the redox-sensitive signaling pathways activated by angiotensin II. Role in vascular smooth muscle cell hypertrophy. J Biol Chem 273(24):15022–15029
102. Touyz RM, Schiffrin EL (1999) Ang II-stimulated superoxide production is mediated via phospholipase D in human vascular smooth muscle cells. Hypertension 34(4 Pt 2):976–982
103. Schieffer B, Luchtefeld M, Braun S, Hilfiker A, Hilfiker-Kleiner D, Drexler H (2000) Role of NAD(P)H oxidase in angiotensin II-induced JAK/STAT signaling and cytokine induction. Circ Res 87(12):1195–1201
104. Du J, Peng T, Scheidegger KJ, Delafontaine P (1999) Angiotensin II activation of insulin-like growth factor 1 receptor transcription is mediated by a tyrosine kinase-dependent redox-sensitive mechanism. Arterioscler Thromb Vasc Biol 19(9):2119–2126
105. Modesti A, Bertolozzi I, Gamberi T, Marchetta M, Lumachi C, Coppo M, Moroni F, Toscano T, Lucchese G, Gensini GF, Modesti PA (2005) Hyperglycemia activates JAK2 signaling pathway in human failing myocytes via angiotensin II-mediated oxidative stress. Diabetes 54(2):394–401
106. Griendling KK, Sorescu D, Lassegue B, Ushio-Fukai M (2000) Modulation of protein kinase activity and gene expression by reactive oxygen species and their role in vascular physiology and pathophysiology. Arterioscler Thromb Vasc Biol 20(10):2175–2183
107. Song B, Jin H, Yu X, Zhang Z, Yu H, Ye J, Xu Y, Zhou T, Oudit GY, Ye JY, Chen C, Gao P, Zhu D, Penninger JM, Zhong JC (2013) Angiotensin-converting enzyme 2 attenuates oxidative stress and VSMC proliferation via the JAK2/STAT3/SOCS3 and profilin-1/MAPK signaling pathways. Regul Pept 185:44–51. doi:10.1016/j.regpep.2013.06.007
108. Campbell JH, Fennessy P, Campbell GR (1992) Effect of perindopril on the development of atherosclerosis in the cholesterol-fed rabbit. Clin Exp Pharmacol Physiol Suppl 19:13–17
109. Ambrosioni E, Bacchelli S, Degli Esposti D, Borghi C (1992) ACE-inhibitors and atherosclerosis. Eur J Epidemiol 8(Suppl 1):129–133
110. Chobanian AV, Haudenschild CC, Nickerson C, Hope S (1992) Trandolapril inhibits atherosclerosis in the Watanabe heritable hyperlipidemic rabbit. Hypertension 20(4):473–477
111. Nickenig G, Jung O, Strehlow K, Zolk O, Linz W, Scholkens BA, Bohm M (1997) Hypercholesterolemia is associated with enhanced angiotensin AT1-receptor expression. Am J Phys 272(6 Pt 2):H2701–H2707
112. Daugherty A, Rateri DL, Lu H, Inagami T, Cassis LA (2004) Hypercholesterolemia stimulates angiotensin peptide synthesis and contributes to atherosclerosis through the AT1A receptor. Circulation 110(25):3849–3857. doi:10.1161/01.CIR.0000150540.54220.C4. 01.CIR.0000150540.54220.C4 [pii]
113. Hirano T, Ran J, Adachi M (2006) Opposing actions of angiotensin II type 1 and 2 receptors on plasma cholesterol levels in rats. J Hypertens 24(1):103–108
114. Lu H, Balakrishnan A, Howatt DA, Wu C, Charnigo R, Liau G, Cassis LA, Daugherty A (2012) Comparative effects of different modes of renin angiotensin system inhibition on hypercholesterolaemia-induced atherosclerosis. Br J Pharmacol 165(6):2000–2008. doi:10.1111/j.1476-5381.2011.01712.x
115. Lu H, Rateri DL, Feldman DL, Charnigo RJ Jr, Fukamizu A, Ishida J, Oesterling EG, Cassis LA, Daugherty A (2008) Renin inhibition reduces hypercholesterolemia-induced atherosclerosis in mice. J Clin Invest 118(3):984–993. doi:10.1172/JCI32970
116. Investigators O, Yusuf S, Teo KK, Pogue J, Dyal L, Copland I, Schumacher H, Dagenais G, Sleight P, Anderson C (2008) Telmisartan, ramipril, or both in patients at high risk for vascular events. N Engl J Med 358(15):1547–1559. doi:10.1056/NEJMoa0801317
117. Zulli A, Burrell LM, Widdop RE, Black MJ, Buxton BF, Hare DL (2006) Immunolocalization of ACE2 and AT2 receptors in rabbit atherosclerotic plaques. J Histochem Cytochem 54(2):147–150. doi:10.1369/jhc.5C6782.2005
118. Thomas MC, Pickering RJ, Tsorotes D, Koitka A, Sheehy K, Bernardi S, Toffoli B, Nguyen-Huu TP, Head GA, Fu Y, Chin-Dusting J,

Cooper ME, Tikellis C (2010) Genetic Ace2 deficiency accentuates vascular inflammation and atherosclerosis in the ApoE knockout mouse. Circ Res 107(7):888–897. doi:10.1161/CIRCRESAHA.110.219279

119. Zhang C, Zhao YX, Zhang YH, Zhu L, Deng BP, Zhou ZL, Li SY, Lu XT, Song LL, Lei XM, Tang WB, Wang N, Pan CM, Song HD, Liu CX, Dong B, Zhang Y, Cao Y (2010) Angiotensin-converting enzyme 2 attenuates atherosclerotic lesions by targeting vascular cells. Proc Natl Acad Sci U S A 107(36):15886–15891. doi:10.1073/pnas.1001253107
120. Tesanovic S, Vinh A, Gaspari TA, Casley D, Widdop RE (2010) Vasoprotective and atheroprotective effects of angiotensin (1-7) in apolipoprotein E-deficient mice. Arterioscler Thromb Vasc Biol 30(8):1606–1613. doi:10.1161/ATVBAHA.110.204453. ATVBAHA.110.204453 [pii]
121. Pernomian L, do Prado AF, Gomes MS, Pernomian L, da Silva CH, Gerlach RF, de Oliveira AM (2015) MAS receptors mediate vasoprotective and atheroprotective effects of candesartan upon the recovery of vascular angiotensin-converting enzyme 2-angiotensin-(1-7)-MAS axis functionality. Eur J Pharmacol 764:173–188. doi:10.1016/j.ejphar.2015.07.007
122. Habiyakare B, Alsaadon H, Mathai ML, Hayes A, Zulli A (2014) Reduction of angiotensin A and alamandine vasoactivity in the rabbit model of atherogenesis: differential effects of alamandine and Ang(1-7). Int J Exp Pathol 95(4):290–295. doi:10.1111/iep.12087
123. Nehme A, Marcelo P, Nasser R, Kobeissy F, Bricca G, Zibara K (2016) The kinetics of angiotensin-I metabolism in human carotid atheroma: an emerging role for angiotensin (1-7). Vasc Pharmacol 85:50–56. doi:10.1016/j.vph.2016.08.001
124. Clancy P, Seto SW, Koblar SA, Golledge J (2013) Role of the angiotensin converting enzyme 1/angiotensin II/angiotensin receptor 1 axis in interstitial collagenase expression in human carotid atheroma. Atherosclerosis 229(2):331–337. doi:10.1016/j.atherosclerosis.2013.05.022
125. Clancy P, Koblar SA, Golledge J (2014) Angiotensin receptor 1 blockade reduces secretion of inflammation associated cytokines from cultured human carotid atheroma and vascular cells in association with reduced extracellular signal regulated kinase expression and activation. Atherosclerosis 236(1):108–115. doi:10.1016/j.atherosclerosis.2014.06.011
126. Clancy P, Koblar S, Golledge J (2016) Involvement of angiotensin II type 1 and 2 receptors in gelatinase regulation in human carotid atheroma in vitro. J Atheroscler Thromb 23(7):773–791. doi:10.5551/jat.31401
127. Daugherty A, Whitman SC (2003) Quantification of atherosclerosis in mice. Methods Mol Biol 209:293–309
128. Azizi M, Rousseau A, Ezan E, Guyene TT, Michelet S, Grognet JM, Lenfant M, Corvol P, Menard J (1996) Acute angiotensin-converting enzyme inhibition increases the plasma level of the natural stem cell regulator N-acetyl-seryl-aspartyl-lysyl-proline. J Clin Invest 97(3):839–844. doi:10.1172/JCI118484
129. Rousseau A, Michaud A, Chauvet MT, Lenfant M, Corvol P (1995) The hemoregulatory peptide N-acetyl-Ser-Asp-Lys-Pro is a natural and specific substrate of the N-terminal active site of human angiotensin-converting enzyme. J Biol Chem 270(8):3656–3661
130. Haznedaroglu IC, Tuncer S, Gursoy M (1996) A local renin-angiotensin system in the bone marrow. Med Hypotheses 46(6):507–510
131. Mrug M, Stopka T, Julian BA, Prchal JF, Prchal JT (1997) Angiotensin II stimulates proliferation of normal early erythroid progenitors. J Clin Invest 100(9):2310–2314. doi:10.1172/JCI119769
132. Haznedaroglu IC, Savas MC, Benekli M (1997) Renin-like activity in leukemic blast cells: an initial clue to a local renin-angiotensin system in the bone marrow. Ann Hematol 75(1–2):69–70
133. Joshi S, Balasubramanian N, Vasam G, Jarajapu YP (2016) Angiotensin converting enzyme versus angiotensin converting enzyme-2 selectivity of MLN-4760 and DX600 in human and murine bone marrow-derived cells. Eur J Pharmacol 774:25–33. doi:10.1016/j.ejphar.2016.01.007
134. Thatcher SE, Gupte M, Hatch N, Cassis LA (2012) Deficiency of ACE2 in bone-marrow-derived cells increases expression of TNF-alpha in adipose stromal cells and augments glucose intolerance in obese C57BL/6 mice. Int J Hypertens 2012:762094. doi:10.1155/2012/762094
135. Strawn WB, Richmond RS, Ann Tallant E, Gallagher PE, Ferrario CM (2004) Renin-angiotensin system expression in rat bone marrow haematopoietic and stromal cells. Br J Haematol 126(1):120–126. doi:10.1111/j.1365-2141.2004.04998.x
136. Gomez RA, Norling LL, Wilfong N, Isakson P, Lynch KR, Hock R, Quesenberry P (1993)

Leukocytes synthesize angiotensinogen. Hypertension 21(4):470–475

137. Rodgers K, Xiong S, DiZerega GS (2003) Effect of angiotensin II and angiotensin(1-7) on hematopoietic recovery after intravenous chemotherapy. Cancer Chemother Pharmacol 51(2):97–106. doi:10.1007/s00280-002-0509-4
138. Rodgers KE, Xiong S, Steer R, diZerega GS (2000) Effect of angiotensin II on hematopoietic progenitor cell proliferation. Stem Cells 18(4):287–294. doi:10.1634/stemcells.18-4-287
139. Gollan F, Richardson E, Goldblatt H (1948) Hypertension in the systemic blood of animals with experimental renal hypertension. J Exp Med 88(4):389–400
140. Goldblatt H (1938) Studies on experimental hypertension: vii. The production of the malignant phase of hypertension. J Exp Med 67(5):809–826
141. Hicks JD, Giltinan P, Pye J (1965) A new method of measuring blood-pressure in mice. Lancet 2(7419):930–932
142. Mills PA, Huetteman DA, Brockway BP, Zwiers LM, Gelsema AJ, Schwartz RS, Kramer K (2000) A new method for measurement of blood pressure, heart rate, and activity in the mouse by radiotelemetry. J Appl Physiol (1985) 88(5):1537–1544
143. Oliverio MI, Kim HS, Ito M, Le T, Audoly L, Best CF, Hiller S, Kluckman K, Maeda N, Smithies O, Coffman TM (1998) Reduced growth, abnormal kidney structure, and type 2 (AT2) angiotensin receptor-mediated blood pressure regulation in mice lacking both AT1A and AT1B receptors for angiotensin II. Proc Natl Acad Sci U S A 95(26):15496–15501
144. Gurley SB, Allred A, Le TH, Griffiths R, Mao L, Philip N, Haystead TA, Donoghue M, Breitbart RE, Acton SL, Rockman HA, Coffman TM (2006) Altered blood pressure responses and normal cardiac phenotype in ACE2-null mice. J Clin Invest 116(8):2218–2225. doi:10.1172/JCI16980
145. Krege JH, John SW, Langenbach LL, Hodgin JB, Hagaman JR, Bachman ES, Jennette JC, O'Brien DA, Smithies O (1995) Male-female differences in fertility and blood pressure in ACE-deficient mice. Nature 375(6527):146–148. doi:10.1038/375146a0
146. Makhanova N, Lee G, Takahashi N, Sequeira Lopez ML, Gomez RA, Kim HS, Smithies O (2006) Kidney function in mice lacking aldosterone. Am J Physiol Renal Physiol 290(1):F61–F69. doi:10.1152/ajprenal.00257.2005
147. Goodwin JE, Zhang J, Geller DS (2008) A critical role for vascular smooth muscle in acute glucocorticoid-induced hypertension. J Am Soc Nephrol 19(7):1291–1299. doi:10.1681/ASN.2007080911
148. Marmot MG (1985) Psychosocial factors and blood pressure. Prev Med 14(4):451–465
149. Morrison JF, Pickford M (1971) Sex differences in the changes in sympathetic nerve activity when arterial pressure is raised by infusion of angiotensin and noradrenaline. J Physiol 216(1):69–85
150. Keys JR, Zhou RH, Harris DM, Druckman CA, Eckhart AD (2005) Vascular smooth muscle overexpression of G protein-coupled receptor kinase 5 elevates blood pressure, which segregates with sex and is dependent on Gi-mediated signaling. Circulation 112(8):1145–1153. doi:10.1161/CIRCULATIONAHA.104.531657
151. Xue B, Pamidimukkala J, Hay M (2005) Sex differences in the development of angiotensin II-induced hypertension in conscious mice. Am J Physiol Heart Circ Physiol 288(5):H2177–H2184. doi:10.1152/ajpheart.00969.2004
152. Venegas-Pont M, Sartori-Valinotti JC, Glover PH, Reckelhoff JF, Ryan MJ (2010) Sexual dimorphism in the blood pressure response to angiotensin II in mice after angiotensin-converting enzyme blockade. Am J Hypertens 23(1):92–96. doi:10.1038/ajh.2009.203
153. Mirabito KM, Hilliard LM, Head GA, Widdop RE, Denton KM (2014) Pressor responsiveness to angiotensin II in female mice is enhanced with age: role of the angiotensin type 2 receptor. Biol Sex Differ 5:13. doi:10.1186/s13293-014-0013-7
154. Gupte M, Thatcher SE, Boustany-Kari CM, Shoemaker R, Yiannikouris F, Zhang X, Karounos M, Cassis LA (2012) Angiotensin converting enzyme 2 contributes to sex differences in the development of obesity hypertension in C57BL/6 mice. Arterioscler Thromb Vasc Biol 32(6):1392–1399. doi:10.1161/atvbaha.112.248559
155. Schneider MP, Wach PF, Durley MK, Pollock JS, Pollock DM (2010) Sex differences in acute ANG II-mediated hemodynamic responses in mice. Am J Physiol Regul Integr Comp Physiol 299(3):R899–R906. doi:10.1152/ajpregu.00638.2009
156. Ji H, Zheng W, Wu X, Liu J, Ecelbarger CM, Watkins R, Arnold AP, Sandberg K (2010) Sex chromosome effects unmasked in angiotensin II-induced hypertension. Hypertension 55(5):1275–1282. doi:10.1161/HYPERTENSIONAHA.109.144949
157. Alsiraj Y, Thatcher SE, Charnigo R, Kuey C, Blalock E, Daugherty A, Cassis LA

(2016) Female mice with an XY sex chromosome complement develop severe angiotensin II-induced abdominal aortic aneurysms. Circulation. doi:10.1161/CIRCULATIONAHA.116.023789

158. Simon MR, Kamlay MT, Khan M, Melmon K (1989) Angiotensin II binding to human mononuclear cells. Immunopharmacol Immunotoxicol 11(1):63–80. doi:10.3109/08923978909082143

159. Luft FC (2001) Angiotensin, inflammation, hypertension, and cardiovascular disease. Curr Hypertens Rep 3(1):61–67

160. Wassmann S, Stumpf M, Strehlow K, Schmid A, Schieffer B, Bohm M, Nickenig G (2004) Interleukin-6 induces oxidative stress and endothelial dysfunction by overexpression of the angiotensin II type 1 receptor. Circ Res 94(4):534–541. doi:10.1161/01.RES.0000115557.25127.8D

161. Touyz RM (2005) Intracellular mechanisms involved in vascular remodelling of resistance arteries in hypertension: role of angiotensin II. Exp Physiol 90(4):449–455. doi:10.1113/expphysiol.2005.030080

162. Guzik TJ, Hoch NE, Brown KA, McCann LA, Rahman A, Dikalov S, Goronzy J, Weyand C, Harrison DG (2007) Role of the T cell in the genesis of angiotensin II induced hypertension and vascular dysfunction. J Exp Med 204(10):2449–2460. doi:10.1084/jem.20070657

163. Yusof M, Kamada K, Gaskin FS, Korthuis RJ (2007) Angiotensin II mediates postischemic leukocyte-endothelial interactions: role of calcitonin gene-related peptide. Am J Physiol Heart Circ Physiol 292(6):H3032–H3037. doi:10.1152/ajpheart.01210.2006

164. Crowley SD, Frey CW, Gould SK, Griffiths R, Ruiz P, Burchette JL, Howell DN, Makhanova N, Yan M, Kim HS, Tharaux PL, Coffman TM (2008) Stimulation of lymphocyte responses by angiotensin II promotes kidney injury in hypertension. Am J Physiol Renal Physiol 295(2):F515–F524. doi:10.1152/ajprenal.00527.2007

165. Shirakawa K, Yan X, Shinmura K, Endo J, Kataoka M, Katsumata Y, Yamamoto T, Anzai A, Isobe S, Yoshida N, Itoh H, Manabe I, Sekai M, Hamazaki Y, Fukuda K, Minato N, Sano M (2016) Obesity accelerates T cell senescence in murine visceral adipose tissue. J Clin Invest. doi:10.1172/JCI88606

166. Marvar PJ, Thabet SR, Guzik TJ, Lob HE, McCann LA, Weyand C, Gordon FJ, Harrison DG (2010) Central and peripheral mechanisms of T-lymphocyte activation and vascular inflammation produced by angiotensin II-induced hypertension. Circ Res 107(2):263–270. doi:10.1161/CIRCRESAHA.110.217299

167. Kintscher U, Hartge M, Hess K, Foryst-Ludwig A, Clemenz M, Wabitsch M, Fischer-Posovszky P, Barth TF, Dragun D, Skurk T, Hauner H, Bluher M, Unger T, Wolf AM, Knippschild U, Hombach V, Marx N (2008) T-lymphocyte infiltration in visceral adipose tissue: a primary event in adipose tissue inflammation and the development of obesity-mediated insulin resistance. Arterioscler Thromb Vasc Biol 28(7):1304–1310. doi:10.1161/ATVBAHA.108.165100

168. Itani HA, Xiao L, Saleh MA, Wu J, Pilkinton MA, Dale BL, Barbaro NR, Foss JD, Kirabo A, Montaniel KR, Norlander AE, Chen W, Sato R, Navar LG, Mallal SA, Madhur MS, Bernstein KE, Harrison DG (2016) CD70 exacerbates blood pressure elevation and renal damage in response to repeated hypertensive stimuli. Circ Res 118(8):1233–1243. doi:10.1161/CIRCRESAHA.115.308111

169. Feau S, Garcia Z, Arens R, Yagita H, Borst J, Schoenberger SP (2012) The CD4(+) T-cell help signal is transmitted from APC to CD8(+) T-cells via CD27-CD70 interactions. Nat Commun 3:948. doi:10.1038/ncomms1948

170. Kirabo A, Fontana V, de Faria AP, Loperena R, Galindo CL, Wu J, Bikineyeva AT, Dikalov S, Xiao L, Chen W, Saleh MA, Trott DW, Itani HA, Vinh A, Amarnath V, Amarnath K, Guzik TJ, Bernstein KE, Shen XZ, Shyr Y, Chen SC, Mernaugh RL, Laffer CL, Elijovich F, Davies SS, Moreno H, Madhur MS, Roberts J 2nd, Harrison DG (2014) DC isoketal-modified proteins activate T cells and promote hypertension. J Clin Invest 124(10):4642–4656. doi:10.1172/JCI74084

171. Saleh MA, McMaster WG, Wu J, Norlander AE, Funt SA, Thabet SR, Kirabo A, Xiao L, Chen W, Itani HA, Michell D, Huan T, Zhang Y, Takaki S, Titze J, Levy D, Harrison DG, Madhur MS (2015) Lymphocyte adaptor protein LNK deficiency exacerbates hypertension and end-organ inflammation. J Clin Invest 125(3):1189–1202. doi:10.1172/JCI76327

172. Briones AM, Nguyen Dinh Cat A, Callera GE, Yogi A, Burger D, He Y, Correa JW, Gagnon AM, Gomez-Sanchez CE, Gomez-Sanchez EP, Sorisky A, Ooi TC, Ruzicka M, Burns KD, Touyz RM (2012) Adipocytes produce aldosterone through calcineurin-dependent signaling pathways: implications in diabetes mellitus-associated obesity and vascular dysfunction.

Hypertension 59(5):1069–1078. doi:10.1161/HYPERTENSIONAHA.111.190223

173. Platten M, Youssef S, Hur EM, Ho PP, Han MH, Lanz TV, Phillips LK, Goldstein MJ, Bhat R, Raine CS, Sobel RA, Steinman L (2009) Blocking angiotensin-converting enzyme induces potent regulatory T cells and modulates TH1- and TH17-mediated autoimmunity. Proc Natl Acad Sci U S A 106(35):14948–14953. doi:10.1073/pnas.0903958106

Chapter 2

A Color Segmentation-Based Method to Quantify Atherosclerotic Lesion Compositions with Immunostaining

Congqing Wu, Alan Daugherty, and Hong Lu

Abstract

There is an increasing recognition that atherosclerotic lesion composition, rather than size, is the determinant of acute events. Immunostaining is a commonly used method to characterize atherosclerotic lesion compositions. Here, we describe a color segmentation-based approach in HSI (hue, saturation, and intensity) color mode, which minimizes subjectivity and produces accurate and consistent quantifications of atherosclerotic lesion compositions.

Key words Immunostaining, Atherosclerosis, Antibody, Imaging, Color

1 Introduction

Mouse models are commonly used to study pathologies and mechanisms of atherosclerosis attributed to the ease of genetic manipulations, availability of large numbers, and cost benefits relative to large species. The two commonly used mouse models, low-density lipoprotein (LDL) receptor −/− and apolipoprotein E (apoE) −/− mice, rapidly develop atherosclerosis when fed a saturated fat-enriched diet (generally referred to as "Western diet") [1–4]. There is also consistent evidence that activation of the renin angiotensin system contributes to hypercholesterolemia-induced atherosclerosis [5–17]. One direct evidence is that infusion of angiotensin II, the major bioactive peptide of the renin angiotensin system, accelerates formation and progression of atherosclerosis in these two hypercholesterolemic mouse models [5, 13, 14]. Conversely, inhibiting key components of the renin angiotensin system reduces atherosclerosis, as demonstrated in both animal models and human trials (a few examples from thousands of publications [8, 10, 12, 15–17]. Therefore, determination of the renin angiotensin components in atherosclerotic lesions may provide mechanistic insights into the development of atherosclerosis.

Sean E. Thatcher (ed.), *The Renin-Angiotensin-Aldosterone System: Methods and Protocols*, Methods in Molecular Biology, vol. 1614, DOI 10.1007/978-1-4939-7030-8_2, © Springer Science+Business Media LLC 2017

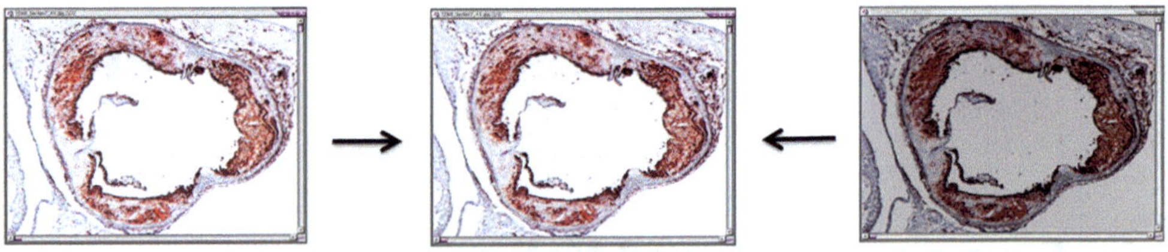

Fig. 1 Normalization of image background. Overexposed or underexposed images can be normalized to same background by the "Best Fit" mode

Immunostaining detects cell types and molecular markers of the pathologic process at different stages of atherosclerotic lesions. This provides insights into the complexity and dynamic nature of lesion development.

Aortic root is a commonly studied region for atherosclerotic lesions in mouse models. Atherosclerotic lesions in mice are characterized by subendothelial accumulation of leukocytes, of which macrophages are the most abundant cell type with lipid content. A cap develops over these macrophages, which contains smooth muscle cells, is also detected in mouse atherosclerotic lesions. Antibodies that specifically target macrophages (e.g., CD68) and the major resident cell type (smooth muscle cells) of the aortic wall have been well established [18]. Therefore, in Subheading 3 method for quantification of positive immunostaining we will use CD68 (Figs. 1, 2, 3, 4, and 5) and smooth muscle α-actin (Fig. 5) positive staining as examples.

Here, we also provide insights into chicken antibodies (Table 1) we developed that target mouse angiotensinogen (AGT), renin, or angiotensin-converting enzyme (ACE), as well as antibodies targeting mouse AngII type 1 (AT1) receptors. Chicken antibodies we developed consist of pre-immune IgY (prior to the introduction of antigen from the same hen for the specific antibody), affinity purified chicken anti mouse antibody, and IgY after eluting the specific antibody (depleted IgY). Therefore, this provides two appropriate negative controls for the antibody, pre-immune IgY and depleted IgY. For renin and ACE immunostaining, we have also used tissues from renin and ACE whole body-deficient mice, respectively, as negative controls to validate the specificity of the antibodies to these two respective targets.

To determine whether these renin angiotensin components are present in mouse atherosclerotic lesions, it is important to first validate these antibodies in their major expressing organs. AGT is ubiquitously expressed in many organs. Hepatocytes are the systemic source for AGT [12, 19, 20]. After synthesis, AGT is released to circulating blood, and also redistributed to other organs such as kidney [12, 21, 22]. Therefore, liver, despite being the major source of systemic AGT, is not optimal to validate an AGT antibody. Instead,

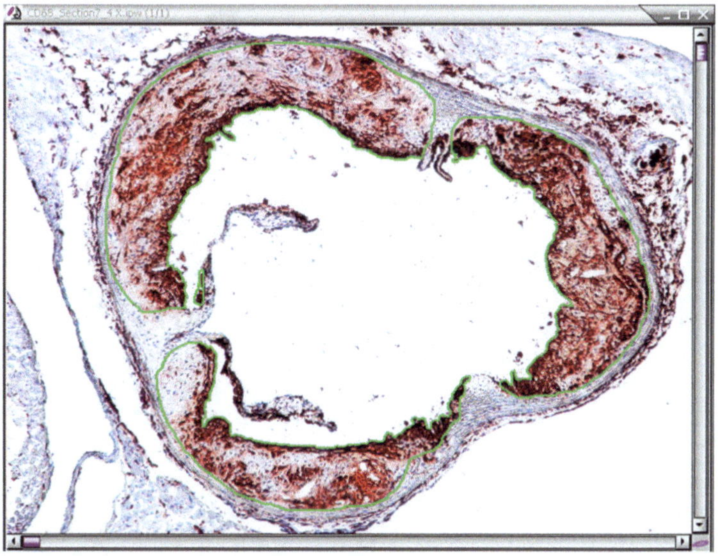

Fig. 2 Defining area of interest. Atherosclerotic lesions in an aortic root section were visualized with CD68 and hematoxylin staining. Individual lesion areas (within green lines) are manually traced using the "Area" tool bar. Three areas of interest are defined as shown by green outlines

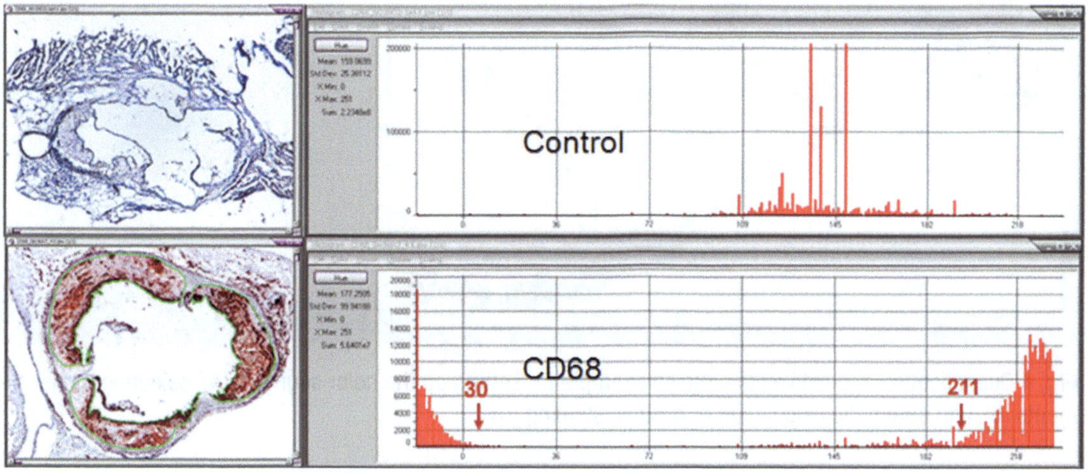

Fig. 3 Determination of color histogram. HUE channels of the color histogram for CD68 positive immunostaining (red) are defined by comparing to the negative control (non-immune rat IgG2a). Two color segments of HUE (0–30) and (211–255) appearing on CD68 positive staining, but not on the negative control, represent CD68 positive immunostaining

AGT accumulates in proximal convoluted tubules of kidney, which provides an optimal tissue for immunostaining of AGT [22]. In contrast to dispersed expression of AGT in many organs, juxtaglomerular cells (JG) of the kidney are the primary source for renin expression. We have demonstrated that chicken anti mouse renin antibody listed

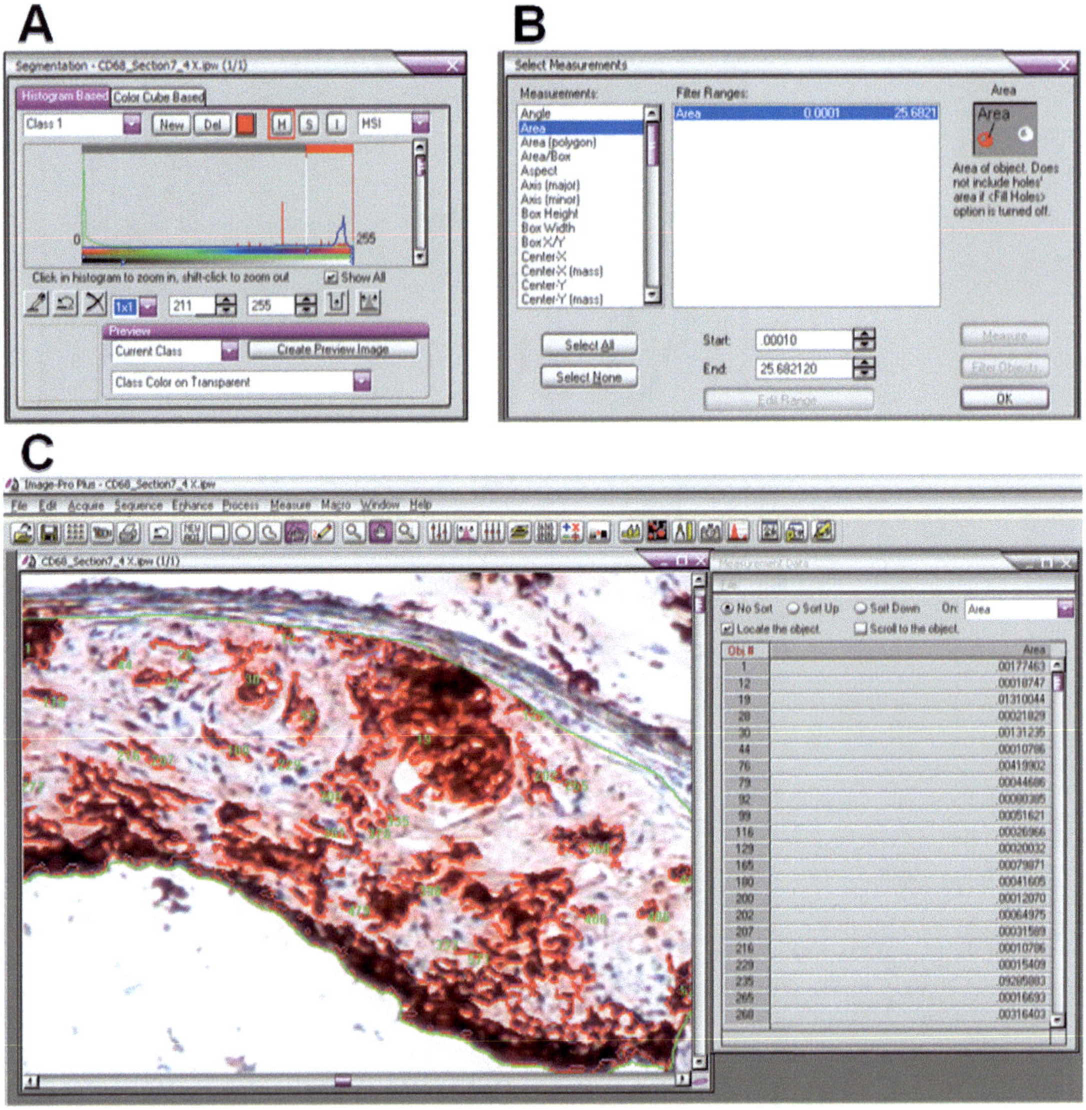

Fig. 4 Quantification of positive staining. An example for setting up the color segment (**A**), selecting measurements (**B**), and automatic calculation of positive staining (**C**)

in Table 1 only detects renin in kidney JG cells of wild-type, but not renin −/− mice. ACE is predominantly expressed in vasculature, which is abundant in both endothelial and smooth muscle cells [6, 11, 23]. Therefore, vessels such as the aorta are appropriate tissues for ACE immunostaining. We have demonstrated that chicken anti mouse ACE antibody listed in Table 1 only develops positive staining in wild-type, but not ACE-deficient tissues.

Angiotensin II contributes to atherosclerosis through its interaction with AT1a receptors [6, 7]. There are many commercially available antibodies against mouse AT1 or AT1a receptors. We have

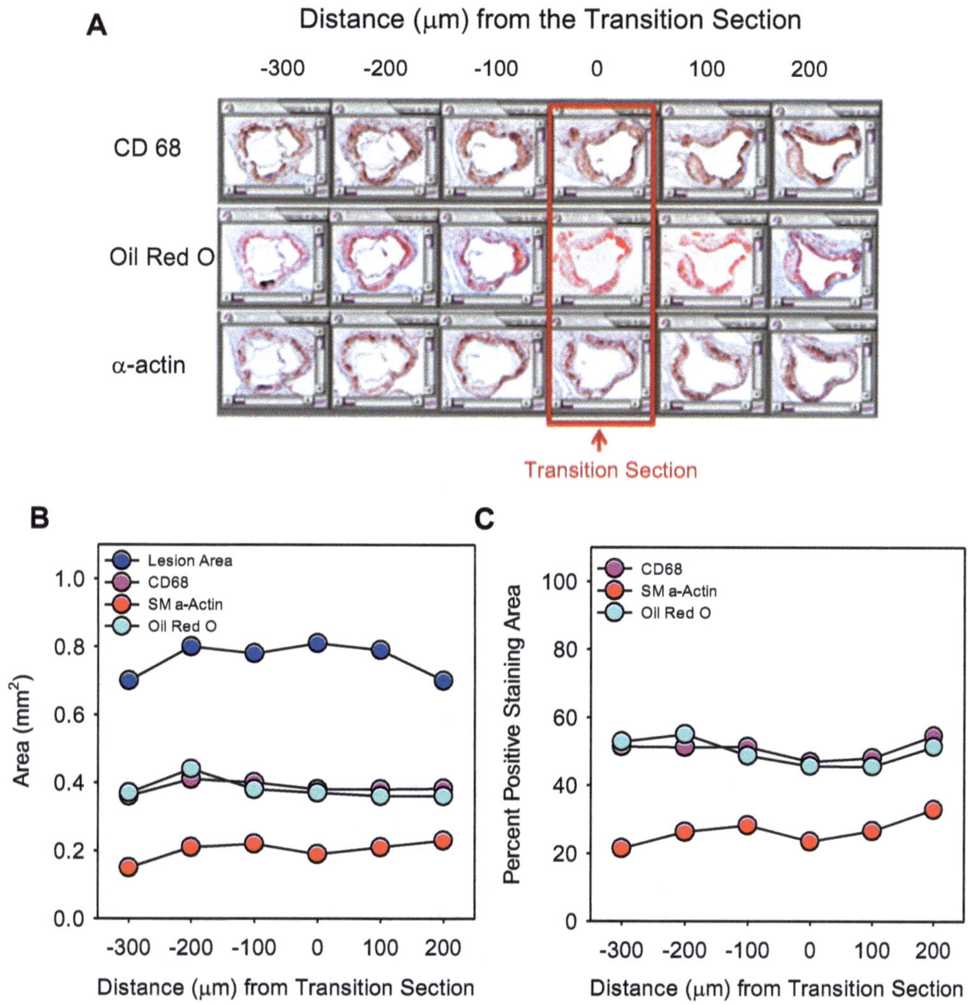

Fig. 5 Analysis of positive staining in atherosclerotic lesions. A male LDL receptor −/− mouse was fed "Western diet" for 7 months. (**A**) Images of CD68, oil *red* O, and smooth muscle α-actin staining (*red* denotes positive staining) throughout the aortic root. "Transition" denotes the ending of the aortic sinus and the beginning of the ascending aorta. (**B**) Positive area of oil *red* O, CD68, and smooth muscle α-actin. (**C**) Percent positive area is calculated by comparing positive staining area of oil *red* O, CD68, and smooth muscle α-actin to lesional area (namely, total area of interest)

Table 1
Chicken anti-mouse IgY antibodies for immunostaining of angiotensinogen, renin, and angiotensin-converting enzyme

Target	Immunogenic peptide sequence	Recommended positive control tissue	Recommended working concentration	References
AGT	CZEEEQPTTSVQQPGSPE	Kidney (proximal convoluted tubules)	10 μg/ml	[6]
Renin	CZRKFYTEFDRHNNR	Kidney (present in JG cells under normal condition)	10 μg/ml	[6, 8]
ACE	CZDLE TDE AKADRFVEEYD RT	Vasculature	3 μg/ml	[6, 11]

AGT angiotensinogen, *ACE* angiotensin-converting enzyme, *JG* juxtaglomerular

also developed multiple chicken anti-mouse antibodies targeting different sequences of the AT1a receptor. Consistent with reports from Dr. Saavedra and Dr. Coffman's laboratories [24, 25], we have failed to demonstrate that AT1 receptor can be stained specifically with any antibody in mouse tissues. Therefore, we are not aware of any antibodies that have validated for AT1a receptor immunostaining of mouse atherosclerosis.

Many image analysis programs provide tools to quantify positive staining with RGB color mode (Red, Green, Blue) as their defaulted setting. While it is optimal to select single homogeneous color [26, 27], RGB mode has shortcomings to quantify compositions on complex immunostaining images. It lacks reproducibility since a range of each color channel is arbitrarily selected by the observer. Therefore, it is difficult to define heterogeneous positive staining colors with RGB color mode, which are commonly seen in atherosclerotic lesions. Here, we describe the use of HSI model for color selection. HSI stands for hue (H), saturation (S), and intensity (I) triplet, each component of which can vary from 0 to 255 in 8-bit computing. Hue defines the color itself. For example, red is distinct from blue and yellow. In theory, the values for the hue axis run from 0 to 360°, beginning and ending with red and running through green, blue, and all intermediary colors such as greenish-blue, orange, purple, and other colors. In an image analysis program, 0–360° is converted to 0–255. Saturation denotes the degree to which the hue differs from a neutral gray. Saturation of 0–100% is converted to 0–255. Intensity means the level of illumination. 0% appears black, whereas 100% is full illumination, which washes out color. The 0–100% mode of intensity is also converted to 0–255. HSI matches the manner how we perceive complex color ranges, while remaining computationally simple. Endpoints of color segments representing positive staining are readily determined without personal bias by comparing histograms between a positive staining image and a negative control image in HSI mode. Appropriately applied, this color segmentation-based method is a reliable technique to provide accurate and consistent quantification of positive immunostaining, a useful tool to look into the development of atherosclerosis as well as many other disease processes.

2 Materials

1. Immunostained cross-sections of aortic root samples.
2. A bright-field microscope with a digital camera for image acquisition.
3. Imaging software that has HSI color mode (Hue, Saturation, Intensity) for quantification of positive immunostaining area and atherosclerotic lesion area.

3 Methods

Macrophages are the most abundant cell type in atherosclerotic lesions of hypercholesterolemic mice, whereas smooth muscle cells are the predominant resident cell type of the aortic wall, which also frequently form a boundary (the cap) of atherosclerotic lesions. Lesions are stained with oil red O to determine the amount of neutral lipid. In this Section, we use immunostaining of CD68 (Clone FA-11) and smooth muscle α-actin as an example to illustrate the Method. Immunostaining of the renin angiotensin components can also be analyzed similarly.

Tissue sections cut on a cryostat were derived from the aortic root of a male LDL receptor−/− mouse fed "Western diet" for 7 months. Sections were stained for neutral lipids with oil red O and immunostained for CD68 (macrophage marker) and smooth muscle α-actin (smooth muscle cell marker). Images of aortic root sections that were 100 μm apart were analyzed using the Image-Pro Plus software. We follow the steps listed below to quantify each image:

1. Image quality control: Make sure all images are taken at same camera settings. Use the "Best Fit" function of the imaging software to fix overexposed or underexposed images (Fig. 1) so that every image has similar background color distribution.
2. Defining area of interest (AOI): Manually trace the whole lesion area as shown in Fig. 2 that green color isolates AOI from the rest of the image. This is to ensure that positive staining is quantified within the defined AOI. Multiple AOIs are allowed. As shown in Fig. 2, three AOIs are defined.
3. Determination of positive immunostaining color range using histogram: Color histogram is a representation of the frequency distribution of colors in an image, determined by counting the number of pixels of each given set of color ranges. This function is under "Measure" in the Image-Pro Plus software. Figure 3 shows the HUE channel of the color histogram of negative control and CD68 positive immunostaining, respectively. Two color segments of HUE (0–30) and (211–255) only appear on CD68 staining but not on its negative control, representing CD68 positive immunostaining. Obtaining these endpoints of color segments using color histogram of the HSI mode is the key to this method (*see* **Note 1**).
4. Quantification of positive immunostaining: After setting color segments with HUE value (Fig. 4a), select "Measurements" as shown in Fig. 4b. Positive staining area is instantly selected and calculated automatically. Selection of positive immunostaining can be modified by adjusting INTENSITY value. The results are shown as presented in Fig. 4c. Use the same HUE and INTENSITY values for every image of same immunostaining (*see* **Note 2**).

5. Analysis of lesional composition: There are two methods to analyze data obtained from **step 4**. One is to determine absolute area of lesional composition as shown in Fig. 5b, and the other is to normalize lesional composition to total lesion area as shown in Fig. 5c. There is no standard which method is better. The selection can depend on what content to state. For example, if the user wants to compare whether there are more macrophages accumulated in lesions in one group than the other, the former method is feasible. If the user wants to emphasize whether the ratio of macrophage accumulation in lesions is more in one group than in the other, the latter method is optimal (*see* **Note 3**).

4 Notes

1. Specificity of positive immunostaining is a premise for accurate quantification [18]. This is determined by appropriate negative controls including antibody-equivalent immunoglobulin from non-immune or pre-immune same host, exclusion of primary antibody, omission of both primary and secondary antibodies. Availability of tissue sections as a negative control would enhance the reliability of the specificity for immunostaining.
2. Quality of images is the key for accurate measurements using this HSI color segmentation-based method. It is also important to acquire all images using same microscope, same camera setting, and same light conditions.
3. Measuring multiple sections (> 5 serial sections), rather than a single section, enhances the accuracy of quantification. Same as atherosclerotic lesion measurements in a specific region [28, 29], this also requires to use a defined landmark to ensure positive immunostaining in same region is compared.

Acknowledgments

Congqing Wu is supported by an American Heart Association Postdoctoral fellow award (16POST31140008). The authors' research work is supported by an Institutional Development Award from the National Institute of General Medical Sciences of the National Institutes of Health under grant number P20 GM103527 and R01 under grant numbers HL107319 and HL133723 from the National Institutes of Health of the United States of America. The content in this manuscript is solely the responsibility of the authors and does not necessarily represent the official views of the National Institutes of Health.

References

1. Breslow JL (1996) Mouse models of atherosclerosis. Science 272:685–688
2. Ishibashi S, Goldstein JL, Brown MS, Herz J, Burns DK (1994) Massive xanthomatosis and atherosclerosis in cholesterol-fed low density lipoprotein receptor-negative mice. J Clin Invest 93:1885–1893
3. Plump AS, Smith JD, Hayek T, Aaltosetala K, Walsh A, Verstuyft JG et al (1992) Severe hypercholesterolemia and atherosclerosis in apolipoprotein-E-deficient mice created by homologous recombination in ES cells. Cell 71:343–353
4. Zhang SH, Reddick RL, Piedrahita JA, Maeda N (1992) Spontaneous hypercholesterolemia and arterial lesions in mice lacking apolipoprotein E. Science 258:468–471
5. Daugherty A, Manning MW, Cassis LA (2000) Angiotensin II promotes atherosclerotic lesions and aneurysms in apolipoprotein E-deficient mice. J Clin Invest 105:1605–1612
6. Daugherty A, Rateri DL, Lu H, Inagami T, Cassis LA (2004) Hypercholesterolemia stimulates angiotensin peptide synthesis and contributes to atherosclerosis through the AT1A receptor. Circulation 110:3849–3857
7. Wassmann S, Czech T, van Eickels M, Fleming I, Bohm M, Nickenig G (2004) Inhibition of diet-induced atherosclerosis and endothelial dysfunction in apolipoprotein E/angiotensin II type 1A receptor double-knockout mice. Circulation 110:3062–3067
8. Lu H, Rateri DL, Feldman DL, Charnigo RJ Jr, Fukamizu A, Ishida J et al (2008) Renin inhibition reduces hypercholesterolemia-induced atherosclerosis in mice. J Clin Invest 118:984–993
9. Daugherty A, Lu H, Rateri DL, Cassis LA (2008) Augmentation of the renin-angiotensin system by hypercholesterolemia promotes vascular diseases. Future Lipidol 3:625–636
10. Lu H, Balakrishnan A, Howatt DA, Wu C, Charnigo R, Liau G et al (2012) Comparative effects of different modes of renin angiotensin system inhibition on hypercholesterolaemia-induced atherosclerosis. Br J Pharmacol 165:2000–2008
11. Chen XC, Lu H, Zhao M, Tashiro K, Cassis LA, Daugherty A (2013) Angiotensin-converting enzyme promotes atherosclerosis through an angiotensin I to angiotensin II pathway involving leukocytes. Arterioscler Thromb Vasc Biol 33:2075–2080
12. Lu H, Wu C, Howatt DA, Balakrishnan A, Moorleghen JJ, Chen X et al (2016) Angiotensinogen exerts effects independent of angiotensin II. Arterioscler Thromb Vasc Biol 36:256–265
13. Daugherty A, Cassis L (1999) Chronic angiotensin II infusion promotes atherogenesis in low density lipoprotein receptor -/- mice. Ann N Y Acad Sci 892:108–118
14. Weiss D, Kools JJ, Taylor WR (2001) Angiotensin II-induced hypertension accelerates the development of atherosclerosis in apoE-deficient mice. Circulation 103:448–454
15. Imanishi T, Tsujioka H, Ikejima H, Kuroi A, Takarada S, Kitabata H et al (2008) Renin inhibitor aliskiren improves impaired nitric oxide bioavailability and protects against atherosclerotic changes. Hypertension 52:563–572
16. Yusuf S, Sleight P, Pogue J, Bosch J, Davies R, Dagenais G (2000) Effects of an angiotensin-converting-enzyme inhibitor, ramipril, on cardiovascular events in high-risk patients. The heart outcomes prevention evaluation study investigators. N Engl J Med 342:145–153
17. ONTARGET Investigators, Yusuf S, Teo KK, Pogue J, Dyal L, Copland I et al (2008) Telmisartan, ramipril, or both in patients at high risk for vascular events. N Engl J Med 358:1547–1559
18. Lu H, Rateri DL, Daugherty A (2007) Immunostaining of mouse atherosclerosis lesions. Methods Mol Med 139:77–94
19. Wu C, Xu Y, Lu H, Howatt DA, Balakrishnan A, Moorleghen JJ et al (2015) Cys18-Cys137 disulfide bond in mouse angiotensinogen does not affect AngII-dependent functions in vivo. Hypertension 65:800–805
20. Yiannikouris F, Wang Y, Shoemaker R, Larian N, Thompson J, English VL et al (2015) Deficiency of angiotensinogen in hepatocytes markedly decreases blood pressure in lean and obese male mice. Hypertension 66:836–842
21. Wu C, Lu H, Cassis LA, Daugherty A (2011) Molecular and pathophysiological features of angiotensinogen: a mini review. N Am J Med Sci (Boston) 4:183–190
22. Matsusaka T, Niimura F, Shimizu A, Pastan I, Saito A, Kobori H et al (2012) Liver angiotensinogen is the primary source of renal angiotensin II. J Am Soc Nephrol 23:1181–1189
23. Chen XC, Howatt DA, Balakrishnan A, Moorleghen JJ, Wu CQ, Cassis LA et al (2016) Angiotensin-converting enzyme in smooth muscle cells promotes atherosclerosis. Arterioscler Thromb Vasc Biol 36:1085–1089
24. Benicky J, Hafko R, Sanchez-Lemus E, Aguilera G, Saavedra JM (2012) Six commercially available

angiotensin II AT(1) receptor antibodies are non-specific. Cell Mol Neurobiol 32:1353–1365

25. Herrera M, Sparks MA, Alfonso-Pecchio AR, Harrison-Bernard LM, Coffman TM (2013) Response to lack of specificity of commercial antibodies leads to misidentification of angiotensin type-1 receptor protein. Hypertension 61:e32
26. Zhang G, Chen Y, BilalWaqar A, Han L, Jia M, Xu C et al (2015) Quantitative analysis of rabbit coronary atherosclerosis. Practical techniques utilizing open-source software. Anal Quant Cytol Histol 37:115–122
27. Vrekoussis T, Chaniotis V, Navrozoglou I, Dousias V, Pavlakis K, Stathopoulos EN et al (2009) Image analysis of breast cancer immunohistochemistry-stained sections using ImageJ: an RGB-based model. Anticancer Res 29:4995–4998
28. Daugherty A, Whitman SC (2003) Quantification of atherosclerosis in mice. Methods Mol Biol 209:293–309
29. Daugherty A, Lu H, Howatt DA, Rateri DL (2009) Modes of defining atherosclerosis in mouse models: relative merits and evolving standards. Methods Mol Biol 573:1–15

Chapter 3

Assessment of Protein Carbonylation and Protein Tyrosine Phosphatase (PTP) Oxidation in Vascular Smooth Muscle Cells (VSMCs) Using Immunoblotting Approaches

Sofia Tsiropoulou and Rhian M. Touyz

Abstract

Post-translational modification of proteins, such as phosphorylation and oxidation, plays a major role in cellular signaling by influencing protein structure and function. In vascular cells, in addition to influencing phosphorylation, angiotensin II (Ang II) induces oxidation of proteins, important in redox signaling in the cardiovascular and renal systems. The present chapter describes immunoblotting approaches to assess irreversible protein carbonylation and protein tyrosine phosphatase (PTPs) oxidation status in the proteome of vascular smooth muscle cells (VSMC).

Protein carbonylation is generally measured using the OxyBlot™ approach, whereby derivatization of protein carbonyl groups (C = O) on oxidized amino acids by dinitrophenylhydrazine (DNPH) results in the formation of a stable dinitrophenyl (DNP) hydrazone product. The samples are analyzed by SDS-PAGE and a primary antibody raised against the DNP moiety is used to determine levels of irreversible protein carbonylation in the sample by immunoblotting.

Oxidation of PTPs can be evaluated using a monoclonal antibody against the "hyperoxidized" (SO_3H) catalytic site of these enzymes. The described methodology offers the ability to discriminate between irreversible (SO_3H) and reversible (SOH) PTP oxidation states. Initially, the free unmodified PTP-thiols (S^-) are alkylated and the sample is split into two. One part is used to assess the PTP-SO_3H form. In the other part reversibly modified PTP-thiols are first reduced and then hyperoxidized by pervanadate (PV). Both untreated and PV-treated samples are analyzed by SDS-PAGE and "hyperoxidized" PTPs are detected by immunoblotting. The proportion of reversibly oxidized PTP-SOH fraction is determined by the difference between the signals in untreated and the PV-treated samples.

The above immunoassays provide general approaches to detect and quantify global levels of irreversible protein oxidation and of irreversibly/reversibly oxidized PTPs in any (patho)physiological context. Characterization of the global redox status is essential to better understand the redox-sensitive mechanisms underlying chronic diseases associated with oxidative stress. This is particularly important in systems influenced by the renin angiotensin system, because Ang II is a potent inducer of oxidative stress and redox signaling.

Key words Protein oxidation, Carbonylation, Oxyblot, PTP oxidation, Cysteine thiol, Irreversible modifications, Reversible modifications

Sean E. Thatcher (ed.), *The Renin-Angiotensin-Aldosterone System: Methods and Protocols*, Methods in Molecular Biology, vol. 1614, DOI 10.1007/978-1-4939-7030-8_3, © Springer Science+Business Media LLC 2017

1 Introduction

1.1 Protein Oxidation

Reactive oxygen species (ROS), such as superoxide anion ($O_2^{\cdot -}$) and hydrogen peroxide (H_2O_2), are natural by-products of enzymatic activity, which at physiological low levels play a key role in cell homeostasis by participating in redox signaling and antioxidant defence. However, excess ROS generation in the cellular environment leads to alterations in cellular function and even cell death, through lipid peroxidation, protein oxidation, and DNA damage. Sustained imbalance in ROS concentration in the system is associated with pathological conditions such as hypertension [1]. Angiotensin II (Ang II), a vasoactive peptide and a major pro-hypertensive hormone, has been implicated in processes associated with endothelial dysfunction, arterial remodeling, vascular fibrosis, and inflammation by stimulating generation of ROS [2].

Interaction of protein macromolecules with highly reactive oxygen and nitrogen species (ROS/RNS) or electrophiles leads to oxidative post-translational protein modification. Such modifications alter the biochemical properties of proteins, including its function and activity, structure, localization, and interactions with other macromolecules, resulting in aberrant cell signaling and function [3]. Depending on the degree of oxidation, protein modifications are classified into two categories: (1) irreversible oxidation that eventually leads to loss of function and protein degradation and (2) reversible oxidation, which is emerging as important cellular regulatory mechanism [4, 5].

The different types of oxidative modifications can be assessed by employing different methodologies. Here, we describe immunoblotting approaches for comprehensive detection of irreversible protein carbonylation and oxidation of protein tyrosine phosphatases (PTP), as crucial cell signaling molecules, highly susceptible to Ang II-induced oxidation.

1.2 Immunoblot Detection of Protein Carbonylation

Carbonylation is the most frequent form of irreversible protein oxidation as it occurs on a number of amino acids including arginine, lysine, proline, threonine, histidine, and cysteine. As such, carbonylation is used as a marker of increased oxidative stress [6]. Metal-catalyzed oxidation of proteins leads to formation of carbonyl groups (C = O; aldehydes and ketones) on side chains of amino acids, with consequent loss of function, aggregation, and ultimately degradation of proteins.

The most commonly used methodology to capture carbonylation is the OxyBlot™ assay, which was initially described by Levine et al. (1994) [7]. Protein carbonyl groups (C = O) are indirectly detected by immunoblotting after reaction with 2,4-dinitrophenylhydrazine (DNPH). VSMCs lysate is initially treated with a reducing agent, such as dithiothreitol (DTT), for reduction of reversibly oxidized proteins and for prevention of artifactual oxidation during sample processing.

Subsequent treatment with DNPH derivatizes carbonyl groups on oxidized proteins, forming a stable dinitrophenyl (DNP) hydrazone product. Samples are analyzed by SDS-PAGE and transferred on an immunoblotting membrane, which is then blocked for nonspecific binding. DNP is detected by the use of a specific primary antibody. Use of horseradish peroxidase (HRP)-conjugated secondary antibody against the primary antibody, along with chemiluminescent substrates, allows for visualization and quantification of total protein carbonyl formation. The intensity of the detected bands is proportional to the degree of protein carbonylation in the sample. Along with the derivatized sample, a negative control (sample treated with derivatization control solution) is analyzed to account for any artifactual carbonylation occurring during the sample processing.

The main advantage of the OxyBlot™ methodology is that no expensive equipment is required, and the main limitation is that no information can be gained regarding the specific site of the modification.

1.3 Immunoblot Assessment of Protein Tyrosine Phosphatases (PTPs) Oxidation States

Protein tyrosine phosphatases (PTPs) are important regulatory molecules, which influence numerous cellular signaling pathways. Under normal conditions, a highly conserved free cysteine thiol at their active center is essential for their catalytic activity on protein dephosphorylation. High levels of ROS react readily with the nucleophilic cysteine thiol, to cause PTP oxidation, which renders the enzyme inactive. The reversibility of the oxidative modification is a crucial mechanism for downstream signaling regulation. Cys oxidation to sulfenic acid (PTP-SOH) can be reversed by antioxidants, whereas higher oxidation to sulfinic (PTP-SO_2H) or sulfonic (PTP-SO_3H) acid states is normally irreversible [8]. A number of approaches have been developed for direct or indirect monitoring of PTP oxidation [9].

The approach described in this chapter, for accessing PTP oxidation, is a modified version of the methodology originally proposed by Persson et al. (2005) [10]. Initially, the method exploits the ability of unmodified PTP-thiols (S^-) to be irreversibly blocked by an alkylating agent such as *N*-ethylmaleimide (NEM). This results in the elimination of those PTPs that remain in a reduced state after a stimulus, from the sample. Subsequently, the cell lysate is split into two. One half of the sample is stored away to be used for directly capturing irreversible PTP-SO_3H hyperoxidation levels. In the second half, reversibly modified PTP-SOH levels are indirectly estimated. -SOH is first reduced into -S^- by dithiothreitol (DTT) and then hyperoxidized into –SO_3H using pervanadate (PV). Both the untreated and PV-treated parts of the same sample are analyzed on the same polyacrylamide gel by SDS-PAGE and transferred on an immunoblotting membrane, which is blocked for nonspecific binding. The hyperoxidized state is captured using a mouse-monoclonal antibody specific to the SO_3H-modified catalytic center of PTPs. Subsequent incubation with a horseradish peroxidase (HRP)-conjugated secondary antibody against the primary antibody and use of chemiluminescent substrates facilitate detection of any PTPs in the $-SO_3H$ state. Calculation of PTP-SOH levels is based on the subtraction of the untreated-sample signal from the PV-treated-sample signal.

The advantage of the strategy described in this chapter is that analysis of the two differentially treated parts of the cell lysate on the same gel enables measurement and direct comparison of the irreversibly and reversibly modified oxidation states of PTPs. Such an approach allows for a more comprehensive assessment of the PTP redox status under different (patho)physiological settings. The main limitation is that this methodology does not capture the irreversible PTP-SO_2H oxidative modification, since it cannot be hyperoxidized by pervanadate [11]. Moreover, the ox-PTP antibody is highly specific only for the classical PTP signature motif and does not detect dual specificity PTPs (DUSPs) [12]. Finally, sulfenic acid (−SOH) being the unstable oxidized intermediate, it can rapidly transform into the more stable irreversible sulfinic or sulfonic acids after cell lysis, leading to false-positive results if samples are not processed fresh and fast enough.

In general, to minimize variability in immunodetection assays, use of multiple replicates and appropriate controls is essential for testing the reproducibility and producing high-quality results.

2 Materials

All aqueous solutions are prepared in Milli-Q grade water (ultrapure; ~18 MΩ.cm), unless otherwise stated.

2.1 Angiotensin II-Stimulation of VSMC

1. Cell culture petri dishes 100 mm.
2. Complete growth media: Dulbecco's Modified Eagle's Medium (DMEM) with 4 mM L-glutamine, 1 g/l D-glucose and 110 mg/l sodium pyruvate, supplemented with 10% (v/v) fetal calf serum (FCS) and 5% (v/v) penicillin (100 IU/ml)—streptomycin (100 μg/ml) solution.
3. Starvation media: DMEM with 4 mM L-glutamine, 1 g/L D-glucose and 110 mg/L sodium pyruvate, supplemented with 0.5% (v/v) FCS and 5% (v/v) penicillin (100 IU/ml)—streptomycin (100 μg/ml) solution.
4. Angiotensin II peptide solution: 10^{-5} M.
5. 1× Phosphate-buffered saline (PBS): 137 mM NaCl, 2.7 mM KCl, 10 mM Na_2PO_4, 1.8 mM KH_2PO_4, pH 7.4.

2.2 OxyBlot

1. Lysis buffer: 50 mM $Na_4P_2O_7$, 50 mM NaF, 50 mM NaCl, 5 mM Na_2EDTA, 10 mM HEPES, 0.5% (v/v) triton X-100, pH 7.4, supplemented with 2 mM Na_3VO_4, 1 mM PMSF and 1 μg/ml of aprotinin, leupeptin and pepstatin.
2. Cell scrapers.
3. OxyBlot™ Protein Oxidation Detection Kit (S7150) from Merck Millipore (Billerica, Massachusetts, USA) (*see* **Note 1**).
4. 1.5 ml Eppendorf tubes: two pre-labeled sets for collection of lysate and sample for Western blotting.
5. 96-well plates: to be used for protein quantification assay and OxyBlot assay.
6. Dithiothreitol (DTT): 1 M stock solution (*see* **Note 2**).
7. Sodium dodecyl sulfate (SDS): 12% w/v.
8. Bio-Rad DC Protein Assay kit (Bio-Rad Laboratories, Hercules, CA).
9. Microplate spectrophotometer.

2.3 PTP Oxidation

1. 1× Phosphate-buffered saline (PBS): 137 mM NaCl, 2.7 mM KCl, 10 mM Na_2PO_4, 1.8 mM KH_2PO_4, pH 7.4.
2. Lysis buffer: 50 mM HEPES pH 6.5, 150 mM NaCl, 1% (v/v) Nonidet P-40, and 10% (v/v) glycerol supplemented with 40 μg/ml PMSF, 2 μg/ml pepstatin, 20 μg/ml leupeptin, 20 μg/ml aprotinin, 20 mM NaF and 10 mM β-glycerophosphate (*see* **Note 3**).
3. Vacuum pump and vacuum flask (*see* **Note 4**).
4. Cell scrapers.
5. 1.5 ml Eppendorf tubes: four pre-labeled sets for collection of lysate, sample without DTT/PV treatment, sample treated with DTT, sample treated with PV.
6. *N*-ethylmaleimide (NEM): added to lysis buffer at a final concentration of 10 mM (*see* **Note 5**).
7. Catalase: added to lysis buffer at a final concentration of 100 μg/ml (*see* **Note 6**).
8. Micro Bio-Spin 6 desalt column pre-packed with specially-sized Bio-Gel®polyacrylamide size exclusion gel and pre-hydrated with equilibration buffer 10 mM Tris pH 7.4 (Bio-Rad) (*see* **Note 7**).
9. Dithiothreitol (DTT): 1 M stock solution (*see* **Note 8**).
10. Tube rotating wheel.
11. Bio-Rad DC Protein Assay kit (Bio-Rad Laboratories, Hercules, CA).
12. Spectrophotometer.

13. Sodium orthovanadate (Na_3VO_4): 100 mM stock solution (*see* **Note 9**).
14. HEPES buffer: 20 mM, pH 7.3.
15. H_2O_2: 30% stock solution.

2.4 Western Blotting

1. Electrophoresis and transfer apparatus.
2. SDS-PAGE resolving buffer: 10% (v/v) acrylamide/bis-acrylamide (40%, 37.5:1), 375 mM Tris–HCl pH 8.8, 0.1% (w/v) SDS, 1% (v/v) TEMED, 10% (v/v) ammonium persulfate (APS).
3. SDS-PAGE stacking buffer: 4% (v/v) acrylamide/bis-acrylamide (40%, 37.5:1), 125 mM Tris–HCl pH 6.8, 0.1% (w/v) SDS, 1% (v/v) TEMED, 10% (v/v) APS.
4. Water saturated isopropanol: 80% isopropanol (v/v).
5. Protein Standard Molecular Weight Marker.
6. 6× Laemmli buffer: 260 mM Tris–HCl pH 6.8, 10% (w/v) SDS, 30% (v/v) glycerol, 0.012% (w/v) bromophenol blue, 6% (v/v) β-mercaptoethanol (*see* **Note 10**).
7. Running buffer: 0.1% (w/v) SDS, 25 mM Tris, 192 mM glycine.
8. Transfer buffer: 25 mM Tris, 192 mM glycine, 20% (v/v) methanol.
9. Polyvinylidene difluoride (PVDF) or nitrocellulose membrane.
10. Ponceau S: 0.5% (w/v) Ponceau S in 1% (v/v) acetic acid.
11. Tris-buffered saline/Tween-20 (TBS-T): 140 mM NaCl, 20 mM Tris pH 7.6, 0.05% (v/v) Tween-20.
12. Blocking buffer: 3% (w/v) BSA/TBS-T or nonfat milk (*see* **Note 11**).
13. Antibody-dilution buffer: 1% (w/v) BSA/TBS-T (*see* **Note 12**).
14. Primary Antibody: Rabbit Anti-DNP antibody (provided in OxyBlot kit); Mouse anti-oxidized-PTP (#MAB2844, R&D Systems, Michigan, USA).
15. Secondary Antibody: Goat Anti-Rabbit IgG (HRP-conjugated) antibody (provided in OxyBlot kit); Rabbit Anti-mouse IgG (HRP-conjugated).
16. Enhanced chemiluminescence reagent (ECL).

3 Methods

3.1 Angiotensin II-Stimulation of VSMC

1. VSMCs are maintained in complete growth media (DMEM/10% FBS), in 100 mm cell culture dishes at 37 °C in a 5% CO_2 humidified incubator, until 70–80% confluent (*see* **Note 13**).
2. Growth media is aspirated and replaced by starvation media (DMEM/0.5% FBS) for ~16 h. (*see* **Note 14**).
3. Starvation media is refreshed 30 min prior to Ang II-stimulation and dishes are returned to the incubator.
4. VSMCs are stimulated with Ang II, 10^{-7} M, for the desirable timepoint(s) (*see* **Note 15**).
5. Stimulation is terminated by washing off the media three times with ice-cold PBS (*see* **Note 16**).
6. Excess PBS is removed completely and VSMCs are processed immediately or dishes are stored at −80 °C until use (*see* **Note 17**).

3.2 OxyBlot Assay

3.2.1 Derivatization of Carbonyl Groups

1. VSMCs are scraped in lysis buffer on ice, using a cell scrapper (*see* **Note 18**).
2. The lysate is transferred into labeled eppendorf tubes and incubated on ice for 30 min (*see* **Note 19**).
3. The sample is centrifuged for 10 min at 12,000 × *g* and at 4 °C.
4. The supernatant is transferred to a new tube and the pellet is discarded.
5. The supernatant is then split:
   (a) 5 μl into a 96-well plate to be used for the protein assay (*see* **Note 20**).
   (b) Half of the sample into an eppendorf tube to be used for the OxyBlot assay.
   (c) The remaining material is saved for western blotting and stored at −80 °C.
6. Based on the protein quantification, all samples designated for OxyBlot are diluted down to the same protein concentration, using lysis buffer.
7. DTT is added to a final concentration of 50 mM and the samples are vortexed briefly (*see* **Note 21**).
8. All the following steps are performed at room temperature (Fig. 1).
9. Equal "x" amounts of sample, containing 15–20 μg of protein, are transferred into two separate wells; one will be subjected to the derivatization reaction (DR) and the second will serve as the negative control (NC) (*see* **Note 22**).

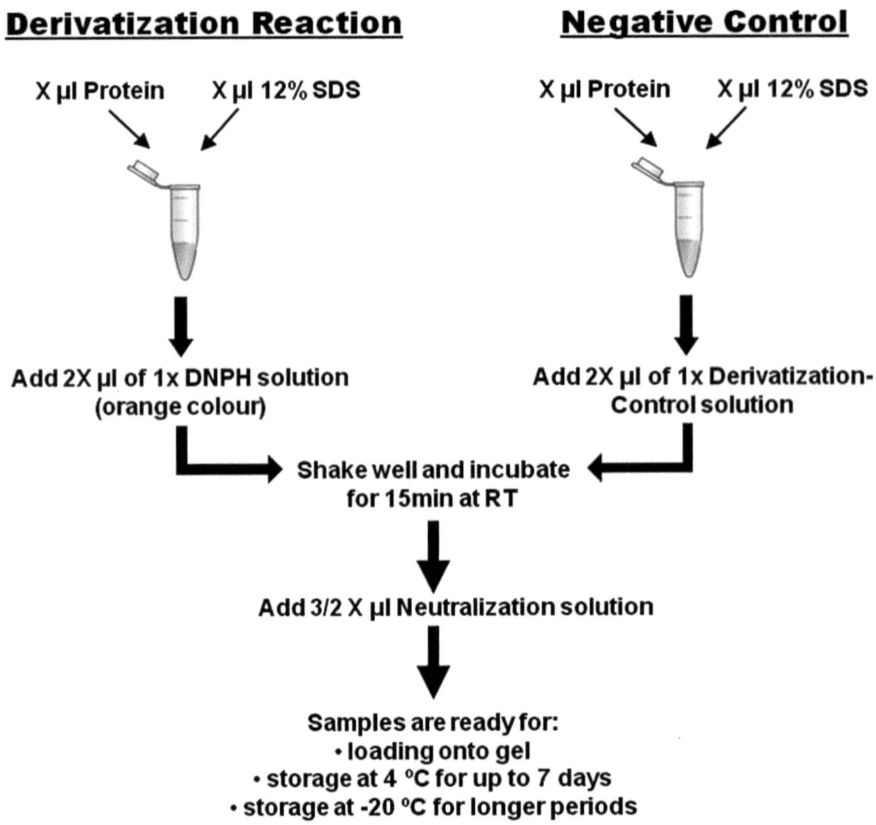

Fig. 1 Oxyblot flowchart. Derivatization of protein carbonyl groups (C = 0) by 2,4-dinitrophenylhydrazine (DNPH)

10. An equal volume "x" of 12% SDS is then added to the sample for protein denaturation.
11. 2 "x" volumes of 1× DNPH or 1× derivatization control solution are added into the wells designated for the DR reaction or NC, respectively (*see* **Note 23**).
12. The samples are mixed by shaking the plate and incubated for 15 min at room temperature (*see* **Note 24**).
13. Derivatization reaction is terminated by addition of 3/2 "x" volumes of neutralization solution to all samples, with sufficient force to mix the sample.
14. Samples can be stored at 4 °C and run on a gel within 7 days. For longer periods of time samples should be aliquoted and stored at −20 °C.

3.2.2 Western Blotting

1. OxyBlot samples must be allowed to reach room temperature prior to loading on a gel (*see* **Note 10**).
2. 10% polyacrylamide gels are casted using the appropriate apparatus and by following the manufacturer's instructions.

3. The samples are loaded in the wells starting with the molecular weight protein standards. For Oxyblot samples equal volumes of each sample are loaded. It is recommended that each DR and the respective NC samples are loaded side by side to allow more efficient comparison during the analysis (*see* **Note 25**).
4. SDS-PAGE is run following standard procedures [13].
5. The proteins are transferred onto a nitrocellulose or PVDF membrane using an appropriate electroblotting apparatus and by following the manufacturer's instructions (*see* **Note 26**).
6. The membrane is soaked into Ponceau S for 5 min and then rinsed with water to visualize the bands and ensure equal loading of the samples.
7. The membrane is rinsed in TBS-T until Ponceau S is washed off.
8. Nonspecific primary antibody binding is blocked by incubating the membrane in blocking/dilution buffer for 1 h at room temperature, under gentle shaking.
9. The membrane is incubated in 15 ml of the primary antibody diluted 1:150 in blocking/dilution buffer, under gentle shaking, overnight, at 4 °C (*see* **Note 27**).
10. Excess, unbound primary antibody is washed off five times for 5 min with TBS-T.
11. The membrane is incubated in 15 ml of the secondary antibody diluted 1:300 in blocking/dilution buffer, for 1 h at room temperature, under gentle shaking.
12. Excess, unbound secondary antibody is washed off three times for 5 min with TBS-T.
13. The membrane is incubated in enhanced chemiluminescence solution according to manufacturer's specifications.
14. The bands are developed either by exposing the membrane to film or to a chemiluminescence western blot imager.

3.2.3 Data Interpretation

1. Signal intensity is directly proportional to the amount of protein carbonyl groups in the samples.
2. Biological samples exposed to stress conditions will have higher carbonylation content.
3. Development of the OxyBlot membrane will detect multiple bands in the derivatization reaction samples and theoretically no bands in the negative controls. Bands appearing in the negative control represent proteins that have undergone oxidation during the sample processing (artifacts) and their signal should be subtracted from the derivatization signal.

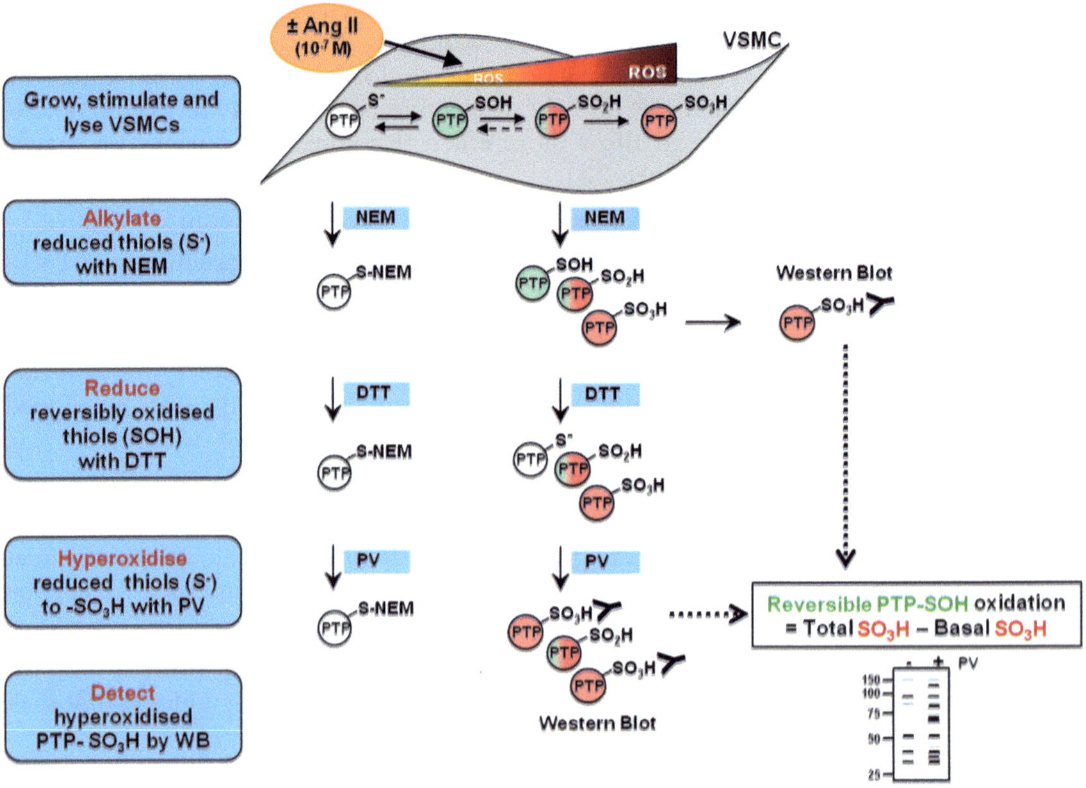

Fig. 2 Flowchart for assessment of PTP oxidation in VSMCs stimulated with Ang II. Bottom right panel, example of oxPTP-immunoblot including samples before (basal $-SO_3H$) and after (total $-SO_3H$) treatment with PV

3.3 PTP Oxidation

3.3.1 Sample Preparation

The following protocol describes a three-step method (S^- alkylation, –SOH reduction, and S^- hyperoxidation) designed to capture both irreversible ($-SO_3H$) and reversible (–SOH) PTP oxidation states (*see* **Note 28**) (Fig. 2).

1. PBS and lysis buffer are degassed for 20 min prior to use, using a vaccum pump (*see* **Notes 4** and **29**).
2. Lysis buffer is supplemented with fresh NEM 10 mM and catalase 100 μg/ml, and degassed for a further 5 min (*see* **Note 5**).
3. VSMCs are rinsed with degassed, ice-cold 1× PBS.
4. Cells are scraped in degassed lysis buffer on ice, using a cell scrapper (*see* **Note 18**).
5. The lysate is transferred into labeled eppendorf tubes and incubated for 1 h at 4 °C, protected from light (*see* **Note 30**).
6. The sample is centrifuged for 15 min at 16,000 × *g* and at 4 °C.

7. The alkylated supernatant is directly transferred onto a Micro Bio-Spin 6 chromatography column (*see* **Note 7**).
8. Excess NEM is removed by centrifugation for 4 min at 1000 × *g* and at 4 °C.
9. The purified flow-through is split into two eppendorf tubes (*see* **Note 31**): One tube will be used for protein quantification and for assessing the irreversible PTP-SO_3H levels and can be stored at −80 °C until use.
10. The flow-through in the other tube is treated with 10 mM DTT for 30 min at room temperature, under rotation, protected from light (*see* **Note 32**).
11. The reduced sample is passed through a Micro Bio-Spin 6 chromatography column.
12. Excess DTT is removed by centrifugation for 4 min at 1000 × *g* and at room temperature.
13. 5 μl of the flow-through is kept for protein quantification.
14. The rest of the purified flow-through is treated with 100 μM PV for 1 h at 4 °C, under rotation, protected from light. For the preparation of 1 mM PV stock solution (*see* **Note 33**):
    (a) 30% H_2O_2 is diluted ten times with HEPES buffer and mixed gently.
    (b) 3% H_2O_2 is further diluted ten times with HEPES buffer and mixed gently.
    (c) 50 μl of 0.3% H_2O_2 is added to 10 μl of 100 mM Na_3VO_4 and 940 μl of H_2O and is mixed gently by reversing the eppendorf tube, to make up 1 mM PV stock.
    (d) After 5 min, a small amount of catalase is scooped using a pipette tip and is mixed into the PV stock, producing a burst of O_2 bubbles. The lid of the tube is left open to release the air pressure (*see* **Note 34**).
    (e) The PV stock solution is good for several hours (<2 h) (*see* **Note 35**).
15. Hyperoxidized samples can be further processed to be analyzed by WB or stored at −80 °C until use.

3.3.2 Western Blotting

1. Equal amounts of protein (20–30 μg) from both DTT/PV untreated and treated samples are diluted to an equal final volume, using lysis buffer free of NEM and catalase.
2. Protein samples are mixed with 6× Laemmli buffer (to 1× final concentration) for denaturation of proteins by β-mercaptoethanol and are boiled at 80 °C, for 10 min.

3. 10% polyacrylamide gels are casted using the appropriate apparatus and by following the manufacturer's instructions.
4. The samples are loaded in the wells starting with the molecular weight protein standards (*see* **Note 36**).
5. Protein electrophoresis, transfer, and membrane blocking are carried out as described in Subheading 3.2.2, **steps 4–8**.
6. The membrane is incubated in 5 ml of the anti-ox-PTP primary antibody at 1 μg/ml blocking/dilution buffer, under gentle shaking, overnight, at 4 °C.
7. Excess, unbound primary antibody is washed off five times for 5 min with TBS-T.
8. The membrane is incubated in 10 ml of the secondary antibody diluted 1:2000 in blocking/dilution buffer, for 1 h at room temperature, under gentle shaking (*see* **Note 37**).
9. Membrane rinsing of the secondary antibody and development is carried out as described in Subheading 3.2.2, **steps 12–14**.

3.3.3 Data Interpretation

1. Signal intensity is directly proportional to the degree of PTP-oxidation in the samples.
2. The cumulative signal of PTP-SOH and PTP-SO_3H oxidation is provided by the samples treated with PV.
3. Levels of irreversible PTP-SO_3H oxidation are provided by those samples not treated with PV.
4. Levels or reversible PTP-SOH oxidation is calculated as the difference between the cumulative and the irreversible PTP-SO_3H oxidation levels.
5. Biological samples exposed to stress-inducing stimuli will exhibit higher PTP-SOH content, whereas PTP-SO_3H content seems to only be affected by exposure to extremely high levels of oxidative stress [12].

4 Notes

1. The mixture of standard proteins with attached DNP residues, provided with the kit, should be aliquoted and stored at −20 °C after the first use. All other reagents are stored at 4 °C.
2. The working concentration of DTT is 50 mM and should not be stored for more than 6 h at room temperature. Prepare just prior to use. Discard any unused reagent.
3. The protease and phosphatase inhibitors are added immediately before use. Sodium orthovanadate (Na_3VO_4) should not be included in the lysis buffer, since it is a known competitive inhibitor of PTPs by occupying their catalytic site, mimicking phospho-Tyrosine, and thus compromising the efficiency of the following steps in this technique [11].

4. Oxygen in the air is dissolved in PBS and lysis buffer and can induce spontaneous PTP oxidation. Therefore, degassing of these solutions prior to use is critical to minimize spontaneous PTP oxidation during cell lysate preparation.
5. NEM is used for free thiol (S^-) alkylation. It is a relatively unstable compound in solution and should be always prepared fresh prior to use. Protect from light by preparing in an amber 1.5 ml tube. Dissolve in a minimum amount of ethanol and then add to lysis buffer.
6. Always add to the lysis buffer fresh, immediately before use.
7. Follow manufacturer's instructions for the preparation of the Micro Bio-Spin 6 chromatography columns before use. Briefly, drain the original storage buffer off the column by short spinning and equilibrate the column by applying degassed lysis buffer three times, before applying the sample. The maximum volume of the sample loaded on the column should not exceed 100 μl.
8. DTT is used for reduction of reversibly oxidized thiols. It is a relatively unstable compound due to air oxidation and solutions should be preferably prepared fresh. Alternatively, stock solution of 1 M can be stored at −20 °C in small aliquots to avoid multiple freezing-thawing cycles.
9. Prepare orthovanadate fresh before use and discard any unused material. The powder must be white/colorless. Yellow coloration means that vanadate (VO_4) forms 12-mer and cannot be used. Alternatively, stock solution of 100 mM can be stored at −20 °C in small aliquots to avoid multiple freezing-thawing cycles.
10. Oxyblot samples do not require mixing with Laemmli buffer and boiling. The neutralization buffer makes the sample dense, so it sinks to the bottom of the well.
11. Both BSA and nonfat milk seem to work equally well with primary antibodies for oxyblot and ox-PTP.
12. Secondary antibodies are normally diluted in 1% (w/v) nonfat milk.
13. In this protocol, primary VSMC were isolated from rat mesenteric bed and used at low-passage (5–6), as healthy cells work better for measuring inducible protein oxidation and respond better to Ang II. Higher passage VSMCs may demonstrate loss of expression of Ang II receptors AT_1R and AT_2R. Growth media is refreshed every 48 h. Cells are split when they reach ~70–80% confluence using trypsin-EDTA for cell detachment and ensuring they were not subject to stress by over or under-confluence. If the experimental design does not require treatment/stimulation of the cells, the complete growth media is replaced by starvation media for approximately 16 h before harvesting.

14. Serum deprivation is used as a technique to induce growth arrest and synchronize all cell cycles to the G_0 quiescent stage before any treatment, so all cells respond in a similar fashion.
15. For VSMCs isolated from rat mesenteric artery the optimal effect of Ang II is achieved at a concentration of 10^{-7} M, according to response curves performed previously in our laboratory. The timepoints used to observe an effect of Ang II on protein/PTP oxidation are relatively short and range between 5 and 60 min. It is important to use negative controls stimulated with just the vehicle, i.e. the solvent that Ang II is dissolved/diluted in, to account for any false-positive effects of the vehicle on the parameters tested. If additional treatment with Ang II-receptor antagonists or antioxidant agents is required, then VSMCs are treated with these agents for a period of 30 min prior to the addition of Ang II. Normally, the final concentration of the antagonists/antioxidants in the cell culture should be at least 10–100 times higher than the stimuli concentration.
16. It is recommended that dishes are kept on ice during the washes to more effectively and acutely stop/slow down any cellular signaling and metabolic reactions, especially if one is interested in capturing very transient changes such as PTP reversible sulfenylation (PTP-SOH).
17. Storage period at −80 °C should not exceed 2–3 months.
18. Harvesting of cells must proceed rapidly to minimize post-lysis spontaneous protein/PTP oxidation due to oxidizing air conditions. The amount of cell lysis buffer used can be adjusted according to the desirable protein concentration of the sample to perform the assays.
19. The samples must be maintained on ice at all times to avoid proteolysis.
20. DTT is not compatible with most protein assays, at the concentration used in this protocol. Therefore, an aliquot of the sample needs to be kept aside for protein quantification, before treatment with DTT.
21. Ideally, OxyBlot should be performed on fresh VSMC lysates. Alternatively, samples treated with DTT can be stored at −20 °C for up to 1 month. DTT is a reducing agent that prevents spontaneous protein oxidation after cell lysis.
22. Larger volumes of sample can be processed to allow running more than one Western blot for carbonylation per sample. Avoid protein concentrations >10 μg/μl, as sample solubility during the derivatization reaction can be affected.
23. The sample for the DR reaction will take an orange color upon addition of DNPH.

24. Samples should not stand in derivatization solution for more than 15–20 min, as this could cause artifact increase in the protein carbonyl content.
25. All samples should have equal protein concentrations, based on the protein quantification assay and the dilutions performed in Subheading 3.2, **step 6**. To further ensure equal loading, probe membranes for GAPDH or any other commonly used loading control.
26. If a PVDF membrane is used, activation of the membrane by soaking it in ethanol prior to use is required.
27. It is recommended to prepare fresh primary antibody each time.
28. If lots of samples/sets need to be analyzed, it is recommended to process fewer samples through the full protocol within 1 day, rather than processing all sets simultaneously but breaking the protocol at more than one step.
29. Lysis buffer may foam due to NP-40. Let the foam sit before use.
30. NEM is light sensitive and is used to alkylate free thiols. It is recommended that a positive control of a sample not treated with NEM is included, to validate the efficiency of the free S^- blocking.
31. The columns can be thoroughly washed with 10 ml of ultrapure H_2O, followed by 10 ml of 20% (v/v) ethanol or PBS supplemented with 0.1% (v/v) sodium azide. Columns can be stored in PBS/sodium azide buffer and can be reused for another three to four times.
32. DTT is light sensitive and is used for reduction/reactivation of the reversibly oxidized PTPs. Omitting this step prevents PV from accessing the active site of reversibly oxidized PTPs and limits hyperoxidation. It is recommended that a negative control of a sample not treated with DTT is included, to evaluate the efficiency of the NEM-mediated alkylation in the first step of the cell lysis.
33. Pervanadate is rapidly converted to vanadate, in the presence of thiol reductants such as DTT [11]. Therefore, complete removal of excess DTT in the previous step is critical for efficient PTP hyperoxidation.
34. Catalase is used to remove excess H_2O_2 that has not reacted with Na_3VO_4 in the solution, seen as a burst of O_2 bubbles.
35. It is recommended to use PV within 5 min to minimize decomposition of the vanadate-H_2O_2 complex.
36. Untreated and PV-treated samples should be loaded on the same gel to allow for direct comparisons between the $–SO_3H$ and $–SOH$ oxidation states of PTPs.
37. Antibody dilutions can be adjusted empirically.

References

1. Montezano AC, Dulak-Lis M, Tsiropoulou S, Harvey A, Briones AM, Touyz RM (2015) Oxidative stress and human hypertension: vascular mechanisms, biomarkers, and novel therapies. Can J Cardiol 31(5):631–641. doi:10.1016/j.cjca.2015.02.008
2. Nguyen Dinh Cat A, Montezano AC, Burger D, Touyz RM (2013) Angiotensin II, NADPH oxidase, and redox signaling in the vasculature. Antioxid Redox Signal 19(10):1110–1120. doi:10.1089/ars.2012.4641
3. Dean RT, Fu S, Stocker R, Davies MJ (1997) Biochemistry and pathology of radical-mediated protein oxidation. Biochem J 324:1–18
4. Cai Z, Yan LJ (2013) Protein oxidative modifications: beneficial roles in disease and health. J Biochem Pharmacol Res 1(1):15–26
5. Tabet F, Schiffrin EL, Callera GE, He Y, Yao G, Ostman A, Kappert K, Tonks NK, Touyz RM (2008) Redox-sensitive signaling by angiotensin II involves oxidative inactivation and blunted phosphorylation of protein tyrosine phosphatase SHP-2 in vascular smooth muscle cells from SHR. Circ Res 103(2):149–158
6. Dalle-Donne I, Giustarini D, Colombo R, Rossi R, Milzani A (2003) Protein carbonylation in human diseases. Trends Mol Med 9:169–176
7. Levine RL, Williams JA, Stadtman ER, Shacter E (1994) Carbonyl assays for determination of oxidatively modified proteins. Methods Enzymol 233:346–357
8. Ostman A, Frijhoff J, Sandin A, Böhmer FD (2011) Regulation of protein tyrosine phosphatases by reversible oxidation. J Biochem 150(4):345–356. doi:10.1093/jb/mvr104
9. Karisch R, Neel BG (2013) Methods to monitor classical protein-tyrosine phosphatase oxidation. FEBS J 280(2):459–475. doi:10.1111/j.1742-4658.2012.08626.x
10. Persson C, Kappert K, Engström U, Ostman A, Sjöblom T (2005) An antibody-based method for monitoring in vivo oxidation of protein tyrosine phosphatases. Methods 35(1):37–43
11. Huyer G, Liu S, Kelly J, Moffat J, Payette P, Kennedy B, Tsaprailis G, Gresser MJ, Ramachandran C (1997) Mechanism of inhibition of protein-tyrosine phosphatases by vanadate and pervanadate. J Biol Chem 272(2):843–851
12. Karisch R, Fernandez M, Taylor P, Virtanen C, St-Germain JR, Jin LL, Harris IS, Mori J, Mak TW, Senis YA, Östman A, Moran MF, Neel BG (2011) Global proteomic assessment of the classical protein-tyrosine phosphatome and "Redoxome". Cell 146(5):826–840. doi:10.1016/j.cell.2011.07.020
13. Laemmli UK (1970) Cleavage of structural proteins during the assembly of the head of bacteriophge T4. Nature 227:680

Chapter 4

Methods for Studying the Role of RAAS in the Modulation of Vascular Repair-Relevant Functions of Stem/Progenitor Cells

Yagna P.R. Jarajapu

Abstract

In recent years, previously unknown functions have been conferred to the RAAS and have been explored in mechanistic studies and disease models. Implication of bone marrow stem/progenitor cells in the cardiovascular protective or detrimental effects of RAAS is a prominent advancement because of the translational significance. Selected members of RAAS are now known to modulate migration, proliferation, and mobilization of bone marrow cells in response to ischemic insult, which are sensitive indicators of vascular repair-relevant functions. In this Chapter, protocols for most frequently used, in vitro, ex vivo, and in vivo assays to explore the potential of RAAS members to stimulate vascular repair-relevant functions of bone marrow stem/progenitor cells of human and murine origin.

Key words $CD34^+$ cells, LSK cells, Bone marrow, Mobilization, Flow cytometry, Migration, Proliferation

1 Introduction

The existence of stem/progenitor cells that accelerate ischemic vascular repair by re-endothelialization has now been well documented in experimental studies. These cells proliferate in response to ischemia-regulated factors, and have the ability to mobilize from the bone marrow niches into systemic circulation and to migrate to the areas of ischemia [1]. Importantly, clinical studies have confirmed the therapeutic potential of progenitor cells for the treatment of ischemic cardiovascular disease.

Evidence has now been accumulated that implicates the renin-angiotensin-aldosterone system (RAAS) in cardiovascular repair and regeneration by modifying the vasoreparative functions of bone marrow stem/progenitor cells [2, 3]. Impaired proliferation or migration in response to stromal-derived factor-1α (SDF) or vascular endothelial growth factor (VEGF) predicts the defective mobili-

Sean E. Thatcher (ed.), *The Renin-Angiotensin-Aldosterone System: Methods and Protocols*, Methods in Molecular Biology, vol. 1614, DOI 10.1007/978-1-4939-7030-8_4,

zation in response to injury and impaired reparative propensity of stem/progenitor cells, all of which are apparent in clinical conditions such as diabetes [4, 5]. We have been consistently using proliferation, migration, and ischemia-induced mobilization as sensitive indicators of vascular repair-relevant functions of human and murine cells for evaluating the potential of ACE2/Angiotensin-(1-7)/Mas receptor pathway [6–10]. Here, we describe preparation of cells from human and mouse samples, and protocols for carrying out these functional assays.

2 Materials

2.1 Cell Culture Reagents

1. C57Bl/6J mice (Jackson laboratories) or C57Bl/6 NHsd (Harlan Laboratories, Inc.).
2. Human peripheral blood, preferably in the form of leukopaks or leucocyte reduction systems (LRS) following apheresis.
3. Buffers/media: PBS (Ca^{2+} and Mg^{2+} free), HBSS, StemSpan®, and RPMI.
4. Stain buffer: PBS with 2% FBS, 0.09% sodium azide.
5. Sorting buffer: PBS with 0.5% FBS.
6. 0.4% Trypan blue solution in PBS, particle-free.
7. 10× RBC lysis buffer.
8. Ficoll® Paque.
9. 96-well plates, U-bottom.
10. 40-μm mesh filters.
11. 5 mL round bottom, and 15 mL and 50 mL conical polypropylene tubes.

2.2 Fluorescence-Activated Cell Sorting (FACS) and Immunomagnetic Enrichment

1. FACSAria cell sorter or FACS jazz for cell sorting and flow cytometry. Accuri C6 flow cytometer.
2. Mouse antibodies: Rat anti-mouse lineage cocktail, anti-Sca-1-APC, and anti-cKit-PE. Rat isotype controls. Mouse anti-human antibodies: CD34-Pacific blue, CD45-PE-Cy7, CD34-PE, CD45-PE-Cy7, and mouse isotype controls.
3. Cell enrichment kits for murine hematopoietic cells, murine $cKit^+$ cells, human hematopoietic cells and human $CD34^+$ cells.

2.3 Assay Kits

1. Colony-forming unit (CFU) assay: Methocult® GF M3434 complete methylcellulose medium with species-specific recombinant cytokines for colony assay.
2. Migration assay: QCM-Chemotaxis-Cell Migration assay, 5 μm, 96-well format, for non-adherent cells.
3. Proliferation assay: Cell proliferation ELISA, BrdU (colorimetric).

3 Methods

In this section, protocols for the isolation of human $CD34^+$ cells or mouse lineage$^-$, c-Kit$^+$ and Sca-1$^+$ cells from peripheral blood or bone marrow by flow-assisted cell sorting or by immunomagnetic enrichment are described followed by functional assays, by which the potential of RAAS components in modulating these functions can be evaluated.

3.1 Low-Density Gradient Separation of Total Mononuclear Cells (MNCs) from Human Peripheral Blood

1. Dilute human blood with PBS if the samples were obtained as a leukopak or in LRS cones with PBS containing 2% FBS in 50 mL tubes and mix gently by inverting tightly capped tubes up and down (*see* **Note 1**).
2. Aliquot 15 mL of Ficoll into 50 mL polypropylene tubes.
3. 35 mL of blood sample is layered over Ficoll (*see* **Note 2**). Mark the cells/Ficoll-paque interface with a felt tip pen.
4. Tubes are centrifuged for 30 min at 800 × *g* at room temperature.
5. The suspension of MNCs is aspirated to the edge of the interface into another 50 mL tube.
6. Cells are slow-washed with PBS to remove platelets and residual plasma and Ficoll by centrifuging at 120 × *g* for 10 min at room temperature.
7. Remove supernatant and resuspend the cell pellet in 45 mL of PBS and repeat **step** 7.
8. Repeat **step 8** for a third wash. Remove supernatant and resuspend pellet in 2 mL PBS.
9. Determine the cell number by diluting an aliquot with Trypan blue solution (1:1) and by using hemocytometer. Trypan blue stains dead cells.

3.2 Enrichment of Lin$^-$ Human Progenitor Cells with CD41 Depletion

This is carried out by using commercially available kits for example from StemCell Technologies.

1. Cell suspension containing 1×10^8 cells/mL, not more than 2 mL, is aliquoted into 5 mL polypropylene tubes.
2. Enrichment cocktail, 50 μL/mL of cells, is added and the cell suspension is gently mixed and incubated at room temperature for 10 min.
3. Nanoparticle suspension is mixed vigorously by using a pipette at least five times to a uniform suspension (*see* **Note 3**).
4. 50 μL of nanoparticle suspension is added to 1 mL of cell suspension and mixed well, and incubated at room temperature for 15 min.
5. Suspension is remixed thoroughly and volume is made up to 5 mL with PBS.

6. Tube is placed in the EasySep magnet and incubated for 10 min at room temperature.
7. Magnet holding the tube is inverted in one continuous motion to decant the desired fraction into another tube. This is the unlabeled fraction or negative selection, enriched for lineage⁻ cells.
8. Total number of cells is determined as described above. This fraction is used for isolation of CD34⁺ cells by FACS or by immunomagnetic positive selection.

3.3 Enrichment of CD34⁺ Cells by FACS

1. Lin⁻ cell suspension (1×10^6 cells/mL stain buffer) is treated with human FcR blocking reagent.
2. Cells are stained with pacific blue-labeled mouse anti-human CD34 and PE-Cy7-labeled mouse anti-human CD45 antibodies will resolve CD34 and CD45 populations with no fluorescence overlap. Cells are incubated for 45 min at 4 °C.
3. Total volume is made up to 5 mL with PBS and centrifuged at $200 \times g$ for 10 min.
4. Supernatant is discarded and **step 3** is repeated.
5. Cell pellet is resuspended in 1 mL of sort buffer.
6. Cells are stained with 7-aminoactinomycin D (7-AAD) to exclude dead cells before sorting (*see* **Note 4**).
7. A tube with cells stained with isotype control antibodies is prepared as per steps above and used for setting the gates for sorting.
8. $CD45^{low}CD34^{high}$ population is selected for sorting.
9. Cells are collected in a sterile tube with medium for culture.
10. Collected cells are enumerated, and spun down at $200 \times g$ and supernatant is discarded.
11. Cell pellet is resuspended in the medium, with or without cytokines, in 96-well plate U-bottom, at a density of 2×10^4 cells/well in 150 μL.

3.4 Alternatively, Lin⁻ Cells Can Be Enriched for CD34 Cells by Immunomagnetic Selection by Using a Kit. With Practice, this Approach Can Obtain 95% Purity of Isolated Cells

1. The Lin⁻ cell suspension is aliquoted at a density of 2×10^8 cells/mL in PBS with 2% FBS and 1 mM EDTA in 5 mL polypropylene tubes.
2. Positive selection cocktail is added to the cell suspension, 100 μL/mL, and mixed thoroughly. Cells are incubated at room temperature for 15 min.
3. Magnetic nanoparticle suspension is mixed vigorously with a pipette to uniform suspension.
4. Nanoparticle suspension, 50 μL/mL, is added to cells and mixed well. Cell suspension is incubated for 10 min at room temperature.

5. Volume is made to 2.5 mL and mixed well.
6. Tube with diluted cell suspension is placed in the EasySep magnet for 5 min at room temperature.
7. Magnet is inverted in one continuous motion to decant the buffer with unwanted cells.
8. **Step** 7 is repeated three more times by adding 2.5 mL of PBS to the tube with retained cells, mixing well before placing in the magnet.
9. $CD34^+$ cells that are retained in the tube are resuspended in PBS and enumerated.
10. Cells are plated in StemSpan as described in Subheading 3.3.

Shown in Fig. 1 are representative dot plots enumerating the $CD34^+$ cell population in MNCs, Lin^- cells or in the preparation enriched as per the protocol in Subheading 3.4 (Fig. 1).

3.5 Low-Density Gradient Separation of Total Mononuclear Cells from Mouse Bone Marrow

Bone marrow cells are obtained by either flushing the bone marrow or by crunching femorae and tibiae.

1. Total bone marrow obtained by crushed bones derived from one mouse is passed through 40 μm mess filters and suspended in PBS with 2% FBS at 3×10^7 cells/mL.
2. Cell suspension is layered over 3 mL of Ficoll® Paque in 15 mL conical tubes.
3. The tubes are spun down at $500 \times g$ for 30 min at room temperature.
4. The interphase cell layer is aspirated into 50 mL of PBS, and slow-washed by centrifugation at $200 \times g$ for 10 min at room temperature.
5. Cells are counted in a hemocytometer as described above.
6. The cells are resuspended in PBS.

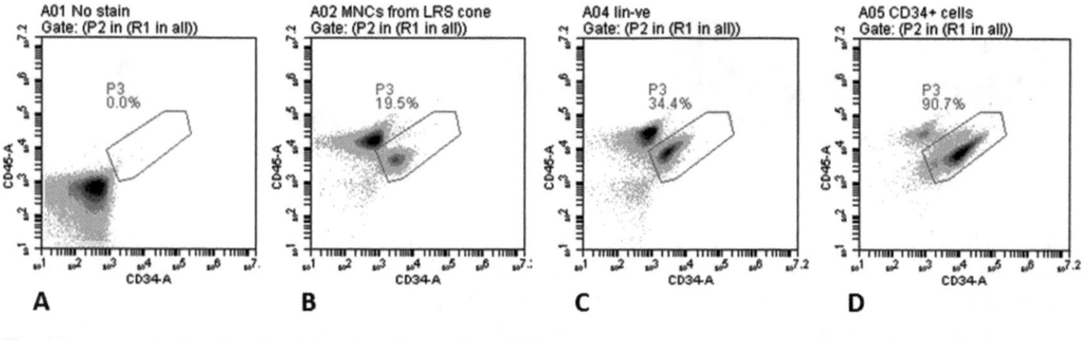

Fig. 1 Representative dot plots showing the purity of $CD34^+$ cells that were enriched by immunomagnetic selection from human peripheral blood. Shown in **a** was an isotype control. Density of $CD34^+$ cells was checked (**b**) in the total mononuclear cells, and (**c**) after enrichment of Lineage negative (Lin^-) cells and (**d**) $CD34^+$ cells

3.6 FACS of Lin^-/$Sca\text{-}1^+$/$c\text{-}Kit^+$ Cells

1. An aliquot of cells in stain buffer with labeled rat isotype antibodies as no-stain control to set the gates for FACS.
2. The rest of cells from Subheading 3.5 are treated with mouse FcR blocking antibody in stain buffer, and stained with lineage cocktail-FITC, c-Kit-PE, and Sca-1-APC.
3. Cell suspensions from **steps 1** and **2** are incubated for 45 min at 4 °C. The tubes are covered with aluminum foil to prevent exposure to light. Concentration of each antibody for mouse cells must be predetermined as per standard protocols.
4. 7-AAD is added before proceeding for sorting to exclude dead cells.
5. Lin^- $Sca\text{-}1^+c\text{-}Kit^+$ cells appear as a well-defined population in a dot plot (Fig. 2).
6. Cells are collected in sterile RPMI and enumerated as described above.
7. Cell suspension is centrifuged at 200 × *g* and the supernatant is discarded.
8. Cells are resuspended in RPMI and plated in a 96-well plate, U-bottom, at a density of 2×10^4 cells/well/150 μL.

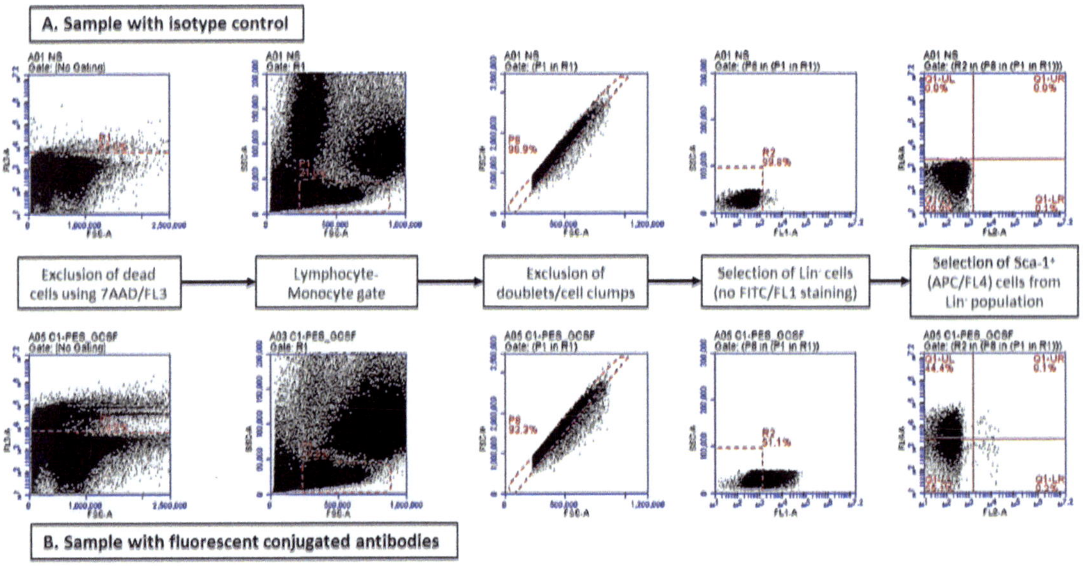

Fig. 2 Representative dot plots showing the protocol for flow cytometric enumeration of lineage-negative (Lin^-), $Sca\text{-}1^+$ and $C\text{-}Kit^+$ cells from mouse peripheral blood. Shown in the top panel is a representative of a sample stained with isotype control antibodies, and shown in the bottom panel resolution of LS, LK, and LSK populations. Note the exclusion of dead cells that are stained by 7-AAD, and the exclusion of doublets or cell clumps by plotting FSC-A against FSC-H

3.7 For Flow Cytometric Enumeration of LSK Cells from Mouse Peripheral Blood or Bone Marrow, the Same Protocol as above Is Followed (Steps 1–5), and Can Be Carried Out in Simple Flow Cytometer with four Different Channels (e.g., Accuri C6)

This is frequently used for evaluating the effect of treatments on mobilization of bone marrow LSK cells by collecting peripheral blood samples at different time intervals.

Shown in Fig. 2 are representative dot plots from the flow cytometric enumeration of LSK cells from peripheral blood.

3.8 Immunomagnetic Enrichment of Mouse Lin⁻ Cells

This is accomplished by using commercially available isolations kits.

1. A suspension of cells obtained in 2.5 is prepared at a density of 1×10^8 cells/mL in Ca^{2+} and Mg^{2+} free PBS with 2% FBS and 1 mM EDTA in a 5 mL polypropylene tube.
2. The cell suspension is treated with naive rat serum at a concentration of 50 μL/mL and mixed thoroughly.
3. Mouse hematopoietic progenitor cell enrichment cocktail is added at a concentration of 50 μL/mL. Suspension is mixed thoroughly and incubated at 2–8 °C for 15 min.
4. Total volume is made up to 2 mL and centrifuged at 200 × *g* for 5 min.
5. Cell pellet is resuspended in the medium as suggested in **step 1**.
6. Biotin selection cocktail is added at a concentration of 100 μL/mL of cell suspension. Suspension is mixed well and incubated at 2–8 °C for 15 min.
7. Magnetic particle suspension is vigorously mixed with a pipette to a uniform suspension.
8. Magnetic particle suspension is added at a concentration of 50 μL/mL. Cell suspension is mixed thoroughly and incubated at 2–8 °C for 10 min.
9. The cell suspension is mixed thoroughly again and volume is made up to 2.5 mL with PBS with 2% FBS and 1 mM EDTA.
10. Cell suspension is then placed in the EasySep magnet and set aside for 3 min.
11. Magnet is inverted in one continuous motion to decant the desired fraction of the cell suspension into a new tube.
12. The number of Lin^- cells collected in **step 11** is determined and are ready for further enrichment.

3.9 Immunomagnetic Enrichment of Mouse $Lin^-c\text{-}Kit^+$ Cells (LK Cells)

Using this approach we can obtain $Lin^-c\text{-}Kit^+$ cells not LSK cells. These cells are preferred over LSK cells for functional assays because of their higher number from bone marrow and these cells possess vascular-repair relevant functions similar to LSK cells. Importantly, a majority of these cells are positive for Sca-1. Along similar lines, LS cells could be isolated in even higher number than LK cells; however, a majority of LS cells are not positive for c-Kit.

1. Lin^- cells obtained in Subheading 3.8 are resuspended in 100 μL of PBS with 2% FBS and 1 mM EDTA in a 5 mL polypropylene tube.
2. Mouse FcR (CD16/32) blocking antibody is added at 1 μg/mL, mixed thoroughly and left aside for 5 min at room temperature.
3. CD117-PE labeling reagent is added 5 μL/100 μL/mL of cell suspension, mixed well and incubated at room temperature for 15 min.
4. PE selection cocktail is then added, 7 μL/100 μL suspension, mixed well and incubated at room temperature for 15 min.
5. Magnetic nanoparticle suspension is vigorously mixed with a pipette to a uniform suspension.
6. Magnetic nanoparticles were added, 5 μL/100 μL suspension, and incubated for 10 min.
7. Cell suspension is made up to 2.5 mL and the tube is kept in the magnet for 5 min.
8. Magnet is inverted in one continuous motion to decant the buffer with unwanted cells.
9. **Step 8** is repeated three more times by adding 2.5 mL of PBS, mixing well before placing in the magnet.
10. LK cells that are retained in the tube are resuspended in PBS and cell count is determined.

3.10 Quantification of Colony Forming Units (CFUs)

This is accomplished by using species-specific medium, Methocult®.

1. Peripheral blood or bone marrow MNCs are mixed with Methocult medium in a 5 mL round-bottom tube.
2. Tubes are vortexed and allowed to rest for 10 min at room temperature.
3. Medium with cells is dispensed into 35 mm culture dishes. Final cell density is 1×10^4 cells/mL for bone marrow cells, or 1×10^5 cells/mL for peripheral blood cells.
4. Dishes are placed in a 100 mm petri dish with an open 35 mm dish with sterile water.
5. The 100 mm dish containing the 35 mm dishes is placed in an incubator at 37 °C, 5% CO_2 for 10 days.

6. CFUs are identified as clusters with more than 50 cells and counted. All colonies, including BFU-E, CFU-GM, and CFU-Mix are counted. Our routine practice is to count the total number of CFUs as a measure of the number of functional progenitor cells present in the given sample.

3.11 Protocol for Migration Assay

This protocol is used for human CD34$^+$ or mouse LK or LSK cells, and is carried out by using QCM-Chemotaxis cell migration assay kit (5 μm) (*see* **Note 5**).

1. Cells are counted and suspended in Hank's buffered saline solution (HBSS). Make sure 2 × 10^4 cells are available for each well/treatment.
2. 150 μL of HBSS is added in the 96-well feeder tray, either control or containing factors such as SDF or VEGF that are species-specific.
3. Cell migration chamber plate, 96-well, is carefully placed on feeder tray and cell suspension containing 2 × 10^4 cells in 100 μL is added to each well.
4. Chemotaxis plate is incubated at 37 °C for 6 h (*see* **Note 6**).
5. The upper migration chamber plate is removed and the cells are discarded by gently flipping off the plate onto blotting paper.
6. Lower chamber with migrated cells is kept at room temperature for use in **step 11**.
7. In another 96-well plate, 150 μL of cell detachment buffer is added, and the migration chamber plate is placed on it and incubated for 30 min at 37 °C.
8. A mixture of cell lysis buffer and a dye solution, provided in the kit, is prepared at 4:1 ratio and dispensed into a black well plate appropriate for fluorescence plate reader.
9. Cell suspension, 75 μL from lower chamber in **step 6** and 75 μL from **step 7** are added to black well plate.
10. Fluorescence is read at 480/520 nm in a plate reader.

    Alternatively, migration assay can be carried out by using CFU assay at **step 4** by using species-specific medium as described in Subheading 3.10.

3.12 Proliferation Assay Protocol

Proliferation of cells can be evaluated in several ways, ranging from simple cell counting or by sequential dilution of intracellular fluorescent dye [9]. We find the assay involving BrdU incorporation is more reproducible, accurate, and efficient as it requires less number of cells per individual treatment. This is carried out by using a colorimetric, cell proliferation BrdU ELISA.

1. Cells are counted and resuspended in appropriate medium, StemSpan for human cells or RPMI for murine cells, with no addition of FBS or supporting factors.

2. Dispense 1 × 10^4 cells per each well/or treatment in 100 μL of medium. Cells with no treatment, control, and cells with 10 μM mitomycin treatment, a negative control, relative to which fold-proliferation will be expressed, are included.
3. Proliferation can be monitored for 24, 48, and 72 h, therefore plan accordingly with required number of cells for the time intervals.
4. Plate with cells and treatments is incubated at 37 °C.
5. Label cells with BrdU for about 18 h with the reagent provided in the kit.
6. Transfer labeled cells from different wells to 1.5 mL eppendorf tubes.
7. Cell suspension is made up to a volume of 1 mL with PBS.
8. Tubes are centrifuged at 120 × *g* for 10 min at room temperature.
9. Supernatant is discarded and dried at 60 °C for 1 h.
10. Fixation-DNAdenat buffer, 200 μL, is added to each tube and tubes are incubated at room temperature for 30 min.
11. Tubes are centrifuged at 120 × *g* for 10 min and the supernatant is discarded.
12. Anti-BrdU-POD, 100 μL/well, is added and then incubated at room temperature for 90 min.
13. Tubes are centrifuged at 120 × *g* for 10 min and the supernatant is discarded.
14. 1 mL of PBS is added and mixed with gentle tapping the bottom of the tube. Tubes are centrifuged at 300 × *g* for 10 min.
15. The supernatant is discarded and the substrate solution is added.
16. Tubes are incubated at room temperature for 15 min.
17. Cells are mixed by pipetting gently for two to three times and then transferred into a 96-well plate (flat-bottom).
18. The absorbance is measured at ODs 405 and 490 nm in a plate reader.
19. A_{405}-A_{490} is an indicator of BrdU incorporation or proliferation.

3.13 Mobilization of Cells in Response to Ischemic Injury

We have standardized this protocol in the model of hind-limb ischemic (HLI) injury in mice [11] and this can be applied to any other models of ischemic insults.

1. Number of mice per group is determined according to the principles of experimental design and power analysis.

2. All mice undergoing ischemic injury are tested for the basal levels of LSK cells in the circulation prior to the ischemic insult. Peripheral blood samples, 70–80 μL, are obtained under isoflurane anesthesia (*see* **Notes 7–9**).
3. Blood samples are treated with RBC lysis buffer, 2–5 mL, in 15 mL tubes, and are incubated at 4–8 °C for 10 min. Volume of RBC lysis is required to be optimized by an individual researcher (*see* **Note 10**).
4. Total volume is made up to10 mL with PBS and centrifuged at 200 × *g* for 10 min.
5. The supernatant is discarded and **step 4** is repeated.
6. If RBCs are still left in the cell pellet, repeat **steps 3–5**.
7. Enumerate the total number of cells as described above.
8. Cells are now ready for enumeration of LSK cells as described in Subheading 3.7.
9. Following ischemic insult, blood samples are collected on days, 1, 2, 3, 5, 7, and 10, and blood samples are processed as described for the numeration of LSK cells (*see* **Note 11**).

4 Notes

1. All centrifugations in the methods described above are preferably carried out in a break-of mode, which would prevent likely disturbance of the cell pellet.
2. Overlaying blood over Ficoll layer is a crucial step that determines clarity of separation of MNC-rich layer from other layers in the gradient. Tilt tube with Ficoll at a 45° angle and overlay blood very slowly by using a 10 mL pipette. An automated pipetting system can be used to overlay blood only if operated in "*g*" (gravity) mode.
3. Thorough mixing of magnetic particles has to be done at every time of use as the particles tend to settle down to the bottom of the vial and form loose clumps relatively quickly.
4. 7-AAD can be conveniently added 5–10 min before flow cytometric enumeration or sorting for staining dead cells at a concentration of 0.25 μg/mL.
5. Migration and proliferation assays can be carried out in human CD34 cells or LK cells obtained from mice that have received pharmacological treatments, or mice that were genetically modified.
6. In our experience with migration assay for non-adherent cells, 5–6 h of incubation is optimal for observing maximum migratory response with standard treatments such as SDF or VEGF

as well as Angiotensin-(1-7). With prolonged periods of time, increased migration is observed; however, the untreated cells also show higher mobilization.

7. Blood sampling for enumeration of circulating bone marrow cells can be done either from orbital sinus or from submandibular vein. The latter is preferred if frequent sampling is needed for example post-HLI for 2–3 weeks at regular time intervals.
8. It is highly recommended that blood sampling for the enumeration of circulating bone marrow cells is performed under light anesthesia. This is necessary to prevent stress-induced mobilization of bone marrow cells, which occurs instantaneously.
9. Blood volume that is required for enumeration of circulating LSK cells requires optimization by an individual researcher and this varies with the treatments or a disease model. For example, following treatment with a strong mobilizer such as granulocyte colony stimulating factor, G-CSF, as low as 25 μL of blood would be more than enough. In contrary, in mice with long-term diabetes, very few cells would be identified in a sample of 100 μL of blood.
10. To lyse RBCs, the recommended volume is in the ratio of 1:4, one volume of cell suspension with four volumes of the lysis buffer, followed by incubation for 10 min at 4 °C. This duration of incubation can be reduced for low volumes of blood; on the other hand for larger volumes of blood this step requires to be repeated. Longer incubation times for larger volumes is not recommended as it would result in damage of cells and poor enrichment. Therefore, this step requires optimization by the individual researcher.
11. In our experience with HLI model in C57Bl/6J mice, peak mobilization of LSK cells is observed in 2–3 days and returns back to basal levels by day 7 following HLI. This pattern, and the number of cells in the circulation at the peak of mobilization, vary with disease, strain, age or following pharmacological treatments that manipulate RAAS. Use of appropriate controls is necessary to avoid false interpretation of results.

Acknowledgments

The author wishes to acknowledge Mr. Shrinidh Joshi, and Mr. Goutham Vasam, for their contributions in refining some of the protocols described in this chapter.

References

1. Asahara T, Kawamoto A, Masuda H (2011) Concise review: circulating endothelial progenitor cells for vascular medicine. Stem Cells 29(11):1650–1655
2. Jarajapu YP, Grant MB, Raizada MK (2012) ACE2/angiotensin-(1-7)/mas axis and cardiovascular regeneration. Curr Hypertens Rev 8(1):35–46
3. Roks AJM, Rodgers K, Walther T (2011) Effects of the renin angiotensin system on vasculogenesis-related progenitor cells. Curr Opin Pharmacol 11:162–174
4. Fadini GP, Derraro F, Quaini F et al (2014) Concise review: diabetes, the bone marrow niche, and impaired vascular regeneration. Stem Cells Transl Med 3(8):949–957
5. Jarajapu YP, Grant MB (2010) The promise of cell based therapies for diabetic complications: challenges and solutions. Circ Res 106(5):854–869
6. Jarajapu YP, Caballero S, Verma A et al (2011) Blockade of NADPH oxidase restores vasoreparative function in diabetic $CD34^+$ cells. Invest Ophthalmol Vis Sci 52(8):5093–5104
7. Jarajapu YP, Bhatwadekar AD, Caballero S et al (2013) Activation of the ACE2/angiotensin-(1-7)/mas receptor axis enhances the reparative function of dysfunctional diabetic endothelial progenitor cells. Diabetes 62(4):1258–1269
8. Singh N, Vasam G, Pawar R et al (2014) Angiotensin-(1-7) reverses angiogenic dysfunction in corpus cavernosum by acting on the microvasculature and bone marrow-derived cells in diabetes. J Sex Med 11(9):2153–2163
9. Jarajapu YP, Hazra S, Segal M et al (2014) Vasoreparative dysfunction of $CD34^+$ cells in diabetic individuals involves hypoxic desensitization and impaired autocrine/paracrine mechanisms. PLoS One 9(4):e93965
10. Singh N, Joshi S, Guo L et al (2015) ACE2/Ang-(1-7)/mas axis stimulates vascular repair-relevant functions of $CD34^+$ cells. Am J Physiol Heart Circ Physiol 309(10):H1687–H1707
11. Niiyama H, Huang NF, Rollins MD et al (2009) Murine model of hindlimb ischemia. J Vis Exp 23:1035. doi:10.3791/1035

Chapter 5

Use of a Fluorescent Substrate to Measure ACE2 Activity in the Mouse Abdominal Aorta

Yu Wang, Lisa A. Cassis, and Sean E. Thatcher

Abstract

The use of fluorogenic substrates to measure enzymatic activity is widely used to understand function within different experimental models. ACE2 is important in understanding the balance between AngII and Ang-(1-7) and how this balance could then in turn influence hypertension or other disease outcomes. Here, we describe a method to measure ACE2 activity in abdominal aorta of hyperlipidemic mice under both saline and AngII infusion.

Key words Angiotensin converting enzyme 2, Aorta, Activity, Enzyme, Fluorescence

1 Introduction

Angiotensin-converting enzyme 2 (ACE2) is a monocarboxypeptidase that cleaves angiotensin II (AngII) to angiotensin-(1-7)(Ang-(1-7)) [1]. ACE2 is known to have other substrates; however, it has been best characterized in the catabolism of AngII [2–4]. Our lab uses iodinated AngII (^{125}I-AngII) and separates out the angiotensin peptides by high-performance liquid chromatography (HPLC) to measure Ang-(1-7) levels in tissues [5, 6]. This method has the benefit of using AngII as the substrate; however, this method requires dedicated lab equipment and the use of radiolabels which may not be practical for all labs. Recently, more labs are using fluorescently quenched compounds to measure enzymatic activity in cells and tissues [7, 8]. This principle uses a quencher molecule, in this case dinitrophenol (Dnp), to block the fluorescence of methoxycoumarin (Mca). This interaction is abolished however when the enzyme cleaves the proline-lysine residue and allows for the Mca to then emit light at a given wavelength. These fluorogenic compounds offer more flexibility since more samples can be screened in a relatively shorter amount of time (e.g., higher throughput). ACE2 is a metalloproteinase; therefore, it requires a

Sean E. Thatcher (ed.), *The Renin-Angiotensin-Aldosterone System: Methods and Protocols*, Methods in Molecular Biology, vol. 1614, DOI 10.1007/978-1-4939-7030-8_5, © Springer Science+Business Media LLC 2017

divalent cation positioned at the active site to perform catalysis. Optimal pH balance is also important for ACE2 and inhibition of angiotensin-converting enzyme 1 (ACE1) and other endopeptidases (e.g., neprilysin and thimet oligopeptidase) are necessary to prevent the influence of these enzymes on Ang-(1-7) formation. Within this protocol, we will use the abdominal aorta as an example of how to measure ACE2 activity in hyperlipidemic mice that have been infused for 28 days with either saline or AngII.

2 Materials

Prepare all solutions using ultrapure water (18 MΩ at 25 °C). Store all reagents at room temperature (unless otherwise indicated) and protect fluorescent substrates from light when working with them at the lab bench. Also, dispose of all wastes properly and according to hazardous waste regulations.

2.1 Stock Solutions

1. 1 M Tris–HCl, pH 7.5: Measure out 75 mL of water and place in beaker with a stir bar. Weigh out 15.76 g of Tris–HCl and place in a beaker and mix to dissolve. Adjust the pH with 1 M NaOH to get to a pH of 7.5. Make up to 100 mL with water.
2. 5 M NaCl solution: Measure out 100 mL of water and place in a beaker with a stir bar. Weigh out 29.2 g of NaCl and place in a beaker and mix to dissolve.
3. 1 M $ZnCl_2$ solution: Measure out 75 mL of water and place in a beaker with a stir bar. Weigh out 13.6 g of $ZnCl_2$ and place in a beaker and mix. Then add 3 mL of 1 N HCl (*see* **Note 1**) to solution and then add water to make up 100 mL.

2.2 ACE1, ACE2, and Prolyl Endopeptidase Inhibitors

1. 100 mM stock of captopril (ACE1 inhibitor): Weigh out a small amount (10–20 milligrams (mg) or 0.01–0.02 g) and divide this amount by 0.02173 g (Molecular weight (MW) of captopril is 217.3) to determine the amount of water needed (*see* **Note 2**). Next, do a tenfold serial dilution twice, to get a final concentration of 1 mM.
2. 10 mM solution of MLN-4760 (ACE2 inhibitor, MW 472.3): Weigh out 4.7 mg and place in 1 mL of water to get this stock solution. Aliquot stocks into eppendorf tubes at 100 μL and store at −20 °C. MLN-4760 is used to determine any activity that is not ACE2 dependent (*see* **Note 3**). You may also use ACE2-deficient cells or tissues as a negative control.
3. 50 mM stock of Z-Pro-Prolinal (Prolyl endopeptidase inhibitor, MW 376.45): Weigh 18.8 mg and dissolve in 1 mL of 50% methanol. Aliquot stocks into eppendorf tubes at 25 μL and store at −20 °C.

4. 1 mg of Mca-APK (Dnp)(ACE2 substrate, MW 696.7): Dilute 1 mg into 143 μL of DMSO to get a 10 mM stock solution (*see* **Note 4**). Store solution at −20 °C and protect from light at all times.
5. Make 10 mL of a 1× ACE2 buffer with the following components: 750 μL of 1 M Tris–HCl, pH 7.5, 2 mL of 5 M NaCl, 5 μL of 1 M $ZnCl_2$, 100 μL of 1 mM captopril, 20 μL of 50 mM Z-Pro-Prolinal, and 7 mL of ultrapure H2O. Make 1 mL of a 10× ACE2 buffer by using 750 μL of 1 M Tris–HCl, pH 7.5, 5 μL of 1 M $ZnCl_2$, and 245 μL of ultrapure H_2O.
6. In order to make up the fluorescent substrate, make the following master mix solution: 265 μL of 1× ACE2 buffer, 10 μL of 10× ACE2 buffer, 20 μL of 5 M NaCl, and 5 μL of fluorescent substrate (10 mM stock, final concentration is 50 μM) (*see* **Note 5**).

3 Methods

Carry out all procedures on ice unless otherwise specified.

1. Add a Complete "Mini" protease inhibitor tablet (EDTA-free) into the 10 mL of 1× ACE2 buffer and vortex until dissolved (*see* **Note 6**).
2. Take 5 mL of the 1× ACE2 buffer and 25 μL of Triton-X to get a 0.5% solution. The other 5 mL of 1× ACE2 buffer will be used for the enzymatic assay.
3. Homogenize the tissue using a Dounce homogenizer (glass-on-glass) in the 0.5% Triton-X 1× ACE2 buffer. For abdominal aorta, we weigh the tissue and use the weight to figure out how much buffer to add (*see* **Note 7**).
4. Centrifuge samples at 12,000 × *g*'s for 15 min at 4 °C. Take clarified supernatant to measure the protein concentration using the Bicinchoninic acid (BCA) assay (*see* **Note 8**). If you have enough protein, perform duplicates for all samples.
5. Take 100 μg of abdominal aortic protein from each sample for loading into the assay (*see* **Note 9**). From the given volume amount subtract it from 75 μL to figure out how much 1× ACE2 buffer to use as a diluent. Load 70 μL of this solution into each well and allow the inhibitors to incubate with the samples for at least 10 min before loading any substrate. Include a "blank" that contains only 1× ACE2 buffer and no protein in your 96-well plate.
6. Before adding the master mix, make sure to turn on spectrophotometer and allow it to warm up for at least 10 min. During this time, you can edit your plate, and set your excitation and emission filters (*see* **Note 10**) (Fig. 1).

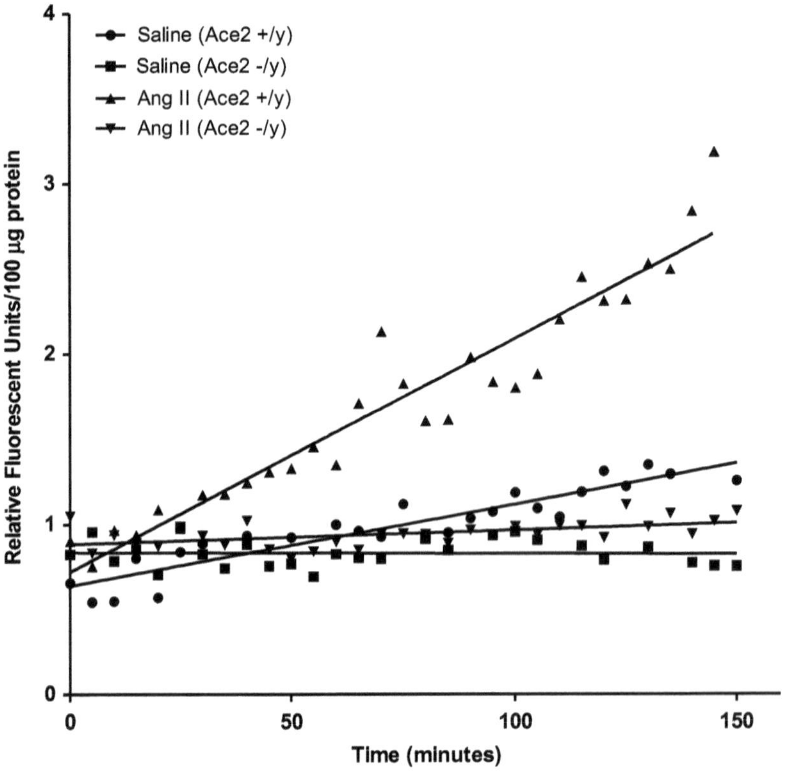

Fig. 1 Abdominal aortic ACE2 activity in saline and angiotensin II (AngII) infused mice (28-day infusion protocol). Note that the ACE2 −/y LDLr−/− mice have slopes that are close to zero, whereas the ACE2 +/y LDLr−/− mice have a slope of 4.8 and a slope of 13.7 (10^{-3} RFU/100 μg protein/5 min) under saline and AngII infusion, respectively. This represents a 2.9-fold increase in ACE2 activity under AngII infusion. Each data point represents an average of 3–5 mice per group

7. Once everything is ready, add 30 μL of master mix using a multi-channel pipettor to all wells and take readings at 5 min for 2.5 h. This will give you a total of 31 readings.

4 Notes

1. A slight amount of hydrochloric acid is needed in order to get $ZnCl_2$ into solution. Heating the solution will not help with solubility of $ZnCl_2$.
2. In order to not waste these compounds, we typically weigh out the smallest amount possible and then use the formula weight to divide to get a certain molarity of the compound. If we weighed out 0.01 g of captopril, then we would need 0.46 mL of water to get a 100 mM stock. Captopril is very stable, so you can keep at room temperature or at 4 °C for 3–6 months.

3. We use MLN-4760 and we incubate tissues or cells with this compound for 30 min at 37 °C before the addition of any ACE2 substrate. There are other ACE2 inhibitors, such as DX-600; however, this inhibitor may have different affinities to human versus mouse ACE2 [7, 8].
4. In our hands, Mca-APK(Dnp) works the best in both ACE2-proficient and ACE2-deficient tissues. There is another fluorescent substrate, Mca-YVADAPK (Dnp); however, this substrate shows a higher background level in ACE2-deficient tissues (unpublished data) and is also reactive to other enzymes, such as caspase-1 and ACE1.
5. You may need to test the amount of substrate needed for your given reactions. We have found that 50 μM works best, however other labs have used less (1–30 μM) [7, 9, 10]. It is important to make sure that the substrate is in excess during the entire incubation time. This master mix is enough for nine reactions. If there are more samples, then you can recalculate for your given number of samples.
6. There are numerous methods for inhibiting proteases and it is important to recognize that the literature has not conformed to one single method. Some labs, including ours, use numerous inhibitors such as pepstatin (renin inhibitor), bestatin (aminopeptidase inhibitor), benzyl succinate (carboxypeptidase inhibitor), and thiorpan/phosphoramadon (neprilysin inhibitors) to inhibit these enzymes. Depending on the tissue type, you may need to test different combinations of inhibitors to find the one that best fits your needs.
7. For abdominal aorta, it is good to use the least amount of buffer necessary to homogenize the tissue. For example, if the abdominal weighs 10 mg, then we will use 100 μL of the homogenization buffer (1 mg per 10 μL of buffer).
8. It should be noted that ACE2 can reside on the cell membrane and within the cell [11]. If cell membrane ACE2 is what is preferred, we would recommend using a high centrifugation rate of 25,000–30,000 × *g*'s at 4 °C for 20 min with 1× ACE2 buffer only without Triton-X. Once you have the pellet, then you resuspend the pellet in 0.5% Triton-X ACE2 buffer. Allow this to incubate overnight and then perform a 5000 × *g*'s spin at 4 °C for 5–10 min to get membrane ACE2. This assay is to look at total ACE2 from abdominal aorta.
9. Abdominal aorta and fat tissue [5, 12] have much lower ACE2 activity than kidney, heart, or lungs; therefore, it is necessary to load a higher amount of protein. It is best to keep this volume as low as possible. The protein lysate should not be any more than 20–25% of the total reaction volume as other proteins could interfere with the assay. It is also wise to try different

amounts of protein lysate to determine what protein concentration will work best for this assay.

10. Make sure that the spectrophotometer can incubate the plate at 37 °C. Also, we do not use the well mixing function before each reading. This can influence the absorbance reading and so the incubation time of 2.5 h may need to be shortened if this function is performed. Also, we do use the auto-cutoff function for performing all spectrophotometry readings at 420 nm. It is best to use a confined emission wavelength as other proteins or substrates could produce background fluorescence within the samples.

References

1. Tipnis SR, Hooper NM, Hyde R, Karran E, Christie G, Turner AJ (2000) A human homolog of angiotensin-converting enzyme. Cloning and functional expression as a captopril-insensitive carboxypeptidase. J Biol Chem 275(43):33238–33243. doi:10.1074/jbc.M002615200. M002615200 [pii]
2. Crackower MA, Sarao R, Oudit GY, Yagil C, Kozieradzki I, Scanga SE, Oliveira-dos-Santos AJ, da Costa J, Zhang L, Pei Y, Scholey J, Ferrario CM, Manoukian AS, Chappell MC, Backx PH, Yagil Y, Penninger JM (2002) Angiotensin-converting enzyme 2 is an essential regulator of heart function. Nature 417(6891):822–828. doi:10.1038/nature00786. nature00786 [pii]
3. Donoghue M, Hsieh F, Baronas E, Godbout K, Gosselin M, Stagliano N, Donovan M, Woolf B, Robison K, Jeyaseelan R, Breitbart RE, Acton S (2000) A novel angiotensin-converting enzyme-related carboxypeptidase (ACE2) converts angiotensin I to angiotensin 1-9. Circ Res 87(5):E1–E9
4. Vickers C, Hales P, Kaushik V, Dick L, Gavin J, Tang J, Godbout K, Parsons T, Baronas E, Hsieh F, Acton S, Patane M, Nichols A, Tummino P (2002) Hydrolysis of biological peptides by human angiotensin-converting enzyme-related carboxypeptidase. J Biol Chem 277(17):14838–14843. doi:10.1074/jbc.M200581200. M200581200 [pii]
5. Gupte M, Boustany-Kari CM, Bharadwaj K, Police S, Thatcher S, Gong MC, English VL, Cassis LA (2008) ACE2 is expressed in mouse adipocytes and regulated by a high-fat diet. Am J Physiol Regul Integr Comp Physiol 295(3):R781–R788. doi:10.1152/ajpregu.00183.2008. 00183.2008 [pii]
6. Thatcher SE, Zhang X, Howatt DA, Lu H, Gurley SB, Daugherty A, Cassis LA (2011) Angiotensin-converting enzyme 2 deficiency in whole body or bone marrow-derived cells increases atherosclerosis in low-density lipoprotein receptor −/− mice. Arterioscler Thromb Vasc Biol 31(4):758–765. doi:10.1161/atvbaha.110.221614
7. Pedersen KB, Sriramula S, Chhabra KH, Xia H, Lazartigues E (2011) Species-specific inhibitor sensitivity of angiotensin-converting enzyme 2 (ACE2) and its implication for ACE2 activity assays. Am J Physiol Regul Integr Comp Physiol 301(5):R1293–R1299. doi:10.1152/ajpregu.00339.2011
8. Ye M, Wysocki J, Gonzalez-Pacheco FR, Salem M, Evora K, Garcia-Halpin L, Poglitsch M, Schuster M, Batlle D (2012) Murine recombinant angiotensin-converting enzyme 2: effect on angiotensin II-dependent hypertension and distinctive angiotensin-converting enzyme 2 inhibitor characteristics on rodent and human angiotensin-converting enzyme 2. Hypertension 60(3):730–740. doi:10.1161/HYPERTENSIONAHA.112.198622
9. Liu J, Ji H, Zheng W, Wu X, Zhu JJ, Arnold AP, Sandberg K (2010) Sex differences in renal angiotensin converting enzyme 2 (ACE2) activity are 17beta-oestradiol-dependent and sex chromosome-independent. Biol Sex Differ 1(1):6. doi:10.1186/2042-6410-1-6
10. Wysocki J, Ye M, Soler MJ, Gurley SB, Xiao HD, Bernstein KE, Coffman TM, Chen S, Batlle D (2006) ACE and ACE2 activity in diabetic mice. Diabetes 55(7):2132–2139. doi:10.2337/db06-0033
11. Lambert DW, Yarski M, Warner FJ, Thornhill P, Parkin ET, Smith AI, Hooper NM, Turner AJ (2005) Tumor necrosis factor-alpha convertase (ADAM17) mediates regulated ectodomain shedding of the severe-acute respiratory syndrome-coronavirus (SARS-CoV) receptor, angiotensin-converting enzyme-2 (ACE2). J Biol Chem 280(34):30113–30119. doi:10.1074/jbc.M505111200. M505111200 [pii]

12. Wang Y, Shoemaker R, Thatcher SE, Batifoulier-Yiannikouris F, English VL, Cassis LA (2015) Administration of 17beta-estradiol to ovariectomized obese female mice reverses obesity-hypertension through an ACE2-dependent mechanism. Am J Physiol Endocrinol Metab 308(12):E1066–E1075. doi:10.1152/ajpendo.00030.2015

Chapter 6

Measuring Blood Pressure Using a Noninvasive Tail Cuff Method in Mice

Yu Wang, Sean E. Thatcher, and Lisa A. Cassis

Abstract

The renin angiotensin system (RAS) is well known for its role in regulating blood pressure (BP). An activated RAS contributes to elevated blood pressure and is evident in both human and animal models of hypertension. Drugs that target the classic vasoconstrictive arm of the RAS (angiotensin II/AT1 receptor signaling) are potent anti-hypertensive agents in clinical setting. However, the newly discovered angiotensin-converting enzyme 2 (ACE2)/angiotensin—(1–7)/Mas receptor axis added new vitality to the hypertension field. Advances in genetic manipulation and the relative low cost made the mouse model as one of the most popular animal models to study hypertension. Since a reliable and accurate method for BP assessment is the key for such experiments, here we provide a protocol for BP measurement in mice using a noninvasive BP system. The CODA noninvasive BP system (a tail-cuff Method, Kent Scientific Corporation) enables blood pressure (BP) measurements in mice. This method uses a specialized volume pressure recording (VPR) sensor, and measures blood volume changes that are placed over the animal's tail. Mice do need to be restrained in specific holders and artificially heated to maintain normal BP.

Key words Blood pressure, Blood volume, Tail cuff, Noninvasive, Mice

1 Introduction

The use of small animal models of hypertension has led to a substantial improvement in understanding the etiology of human hypertension. The ability to accurately assess BP in small animals is the key point to elucidate the underlying determinants of hypertension. Although a direct intra-arterial assessment of BP in unrestrained, conscious animals is generally considered the most translationally relevant, issues associated with this method such as the invasive nature, the technical difficulty of the surgery, and the expensive price limit its usage in a general laboratory setting [1, 2]. Within this protocol, we demonstrate the use of a noninvasive tail-cuff method to measure BP in mice. This tail cuff method is very user friendly and is well correlated with other methods, such as assessment of intra-arterial BP.

Sean E. Thatcher (ed.), *The Renin-Angiotensin-Aldosterone System: Methods and Protocols*, Methods in Molecular Biology, vol. 1614,
DOI 10.1007/978-1-4939-7030-8_6, © Springer Science+Business Media LLC 2017

2 Materials

CODA High Throughput Noninvasive Blood Pressure system (Product # CODA-HT8, Kent Scientific corporation), neutral cleaning reagent and kimwipes.

3 Methods

3.1 Setting Up Equipment and Software

1. The area for setting up the Coda system and performing experiments should be in a room where the temperature is stable (20–23 °C). Avoid locations near air conditioning vents or fans (*see* **Note 1**).
2. Turn on the Coda controller and then use the warming platform (set at appropriate temperature levels accordingly, for room temperature, use level 1 or 2, *see* **Note 2**).
3. Perform the controller diagnostic test.
   - Open Coda software, select coda device by click on it, select (test selected Device), select the cuffs and the channels to test and click (test).
   - After the test's complete, close the device test window. Select the Coda controller to use then click (use these devices).
4. Setting up software (Coda noninvasive blood pressure system).
   - Select (Tools) >> (Manage personnel).
   - Under (Researchers), enter performer's name and save data.
   - Under (Specimens), name animals and animal type and save data.
5. Begin a new experiment.
   - Open (File) >> (New) >> (Experiment).
   - In experiment wizard window, enter experiment name, select key researcher, and begin date and then click (Next) to enter session name information (example: session 1).
   - Set Acclimation Cycles at 5; Number of sets at 1; Time between Sets at 30 s; Cycles per set at 20;
   - Time between Cycles at 5 s, click (Next) to (Specimen selection).
   - Select appropriate specimens from the specimen pool (on left) and assign to available channels (on right) and then click (Next) to (Session Parameters).
   - Set Maximum Occlusion pressure at 250 mmHg; Deflation time at 20 s for mice; Minimum volume at 15 μL, and Display style "One Channel per Graph"; Click (Next) to (Review Session Script) and Click (Next) to (Experiment

Configured). Click (Finish) only after completion of the "Preparing the Animals" Subheading 3.2.

3.2 Preparing the Animals

1. Select appropriate holders for mice based on their size (*see* **Note 3**).
2. Place the animal into a holder (mouse should be able to freely enter the holder) (Fig. 1) [3].
3. Picking up a mouse by the tail and gently place it into the rear of the holder which faces the open end of the nose cone.
4. Carefully secure the rear hatch to the holder by turning a screw on the rear hatch. Be careful not to pinch the tail or any other parts of the body which secure the rear hatch.
5. Next, slide the nose cone toward the rear hatch limiting the movement of the animal. The nose cone should be in a position to limit the animal from turning around while inside the holder.
6. Place the holder onto the warming platform in the designated position. Allow the animal at least 5 min to acclimate to the holder (*see* **Note 4**). Do not touch the animals in the holder, unnecessary contact could irritate the animal. Never leave the animals unattended.
7. Place the occlusion tail cuff through the tail and to the base of tail without force. After that, thread the tail through the VPR sensor cuff and place it within 2 mm of the occlusion cuff (Fig. 2, *see* **Note 5**) [3].
8. Secure the air tubing in the notch on the top rear of the holder.
9. After all the animals have been placed in the holder and their tails cuffed, allow them at least 5 min to acclimate to surroundings. Record the temperature of the animal with an infrared thermometer if needed. The temperature of animals should between 32 and 35 °C.

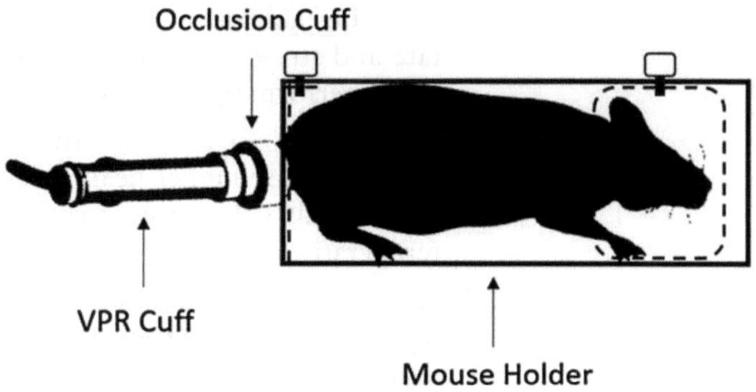

Fig. 1 Schematic of mouse holder for the Kent CODA system. Note the position of the nose cone and rear hatch for proper restraint of the mouse

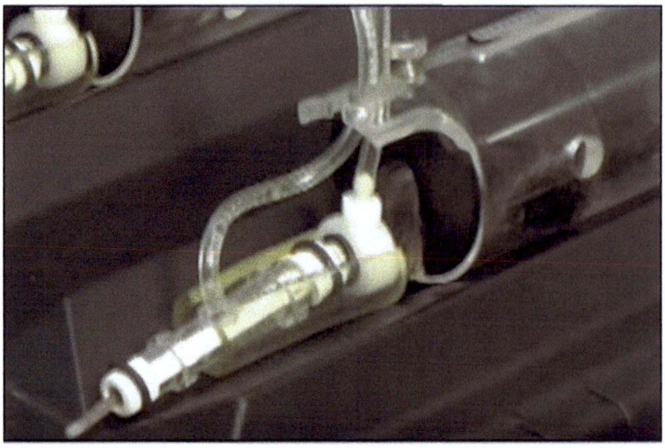

Fig. 2 Proper placement of the occlusion and VPR cuffs on the mouse tail. Note the air tubing is not twisted and placed in open notch of the mouse holder

3.3 Data Collection and Processing

1. Click (Finish) on the (Experiment configured) window to begin the experiment and data collection (*see* **Note 6**).
2. When the experiment has ended, remove the animals immediately from the cuffs and the holder and place back to their own cages.
3. A simple session summary report should be displayed. The data collected in the experiment is saved as an excel file.
4. There are several ways to process the data within the excel file. A common practice is to obtain the average and standard deviation.
5. Delete measurements if the standard deviation is greater than 30. Additionally, rejected cycles may be viewed and the entire data base can be easily exported to excel (*see* **Notes** 7 and **8**).

4 Notes

1. Exposing animals to loud noise or extraneous odors may irritate and stress the animal. Be quite and do not wear perfume when performing experiment.
2. It is critical to maintain the mouse's body temperature at 32–35 °C, use higher heating levels if the room temperature is too low. Also obese mice may take a longer time to reach target temperature.
3. Since BPs were measured by directly sensing changes in volume pressure in the tail with VPR sensor instead of sensing light penetration with a photo-resistor, BP can be measured in dark-skinned mice (C57BL/6J) without difficulty. However, in case of obese mice (usually with thick tail due to excess fat

accumulation underneath the skin), BP measurement using tail cuff method could be challenging and the BP values are usually low.

4. Two to three days of acclimation measurements are usually recommended followed by another two to three days of data collection measurements to assure accurate measurements.
5. The occlusion tail cuff is inflated to inhibit the blood flow to the tail. The occlusion cuff is slowly deflated and the VPR cuff incorporating the specially designed sensor measures the physiological characteristics of the returning blood flow. As the blood returns to the tail, the VPR sensor cuff measures the tail swelling as a result of the arterial pulsations from blood flow. Systolic blood pressure (SBP) is automatically measured at the first appearance of tail swelling and diastolic blood pressure (DBP) is automatically calculated with the increasing rate of swelling in the tail.
6. In the real-time BP graph, the red line represents the deflation of the occlusion cuff and the VPR sensor cuff, the blue line represents blood volume changes in the tail. The first inflection of the blue line, minimum rate of change identifies the SBP; the second inflection of the blue line, maximum rate of change identifies the DBP.
7. Do not schedule the BP measurement in male and female mice at the same time. The presence of an opposite sex may irritate the mice. Always do the male mice first if your experiments happen to have both sexes in the same study design. Clean devices (platforms, holders, etc.) after each round of measurement.
8. Appropriate training for software operating and animal handling is required prior to perform this method for BP measurement. It is most important to reduce animal stress by following this procedure in order to get accurate and consistent measurements.

References

1. Krege JH, Hodgin JB, Hagaman JR, Smithies O (1995) A noninvasive computerized tail-cuff system for measuring blood pressure in mice. Hypertension 25(5):1111–1115
2. Whitesall SE, Hoff JB, Vollmer AP, D'Alecy LG (2004) Comparison of simultaneous measurement of mouse systolic arterial blood pressure by radiotelemetry and tail-cuff methods. Am J Physiol Heart Circ Physiol 286(6):H2408–H2415. doi:10.1152/ajpheart.01089.2003
3. Daugherty A, Rateri D, Hong L, Balakrishnan A (2009) Measuring blood pressure in mice using volume pressure recording, a tail-cuff method. J Vis Exp 27. doi:10.3791/1291

Chapter 7

Blood Pressure Monitoring Using Radio Telemetry Method in Mice

Yu Wang, Sean E. Thatcher, and Lisa A. Cassis

Abstract

The TA11PA-C10 implantable transmitter (Data Sciences International, DSI) is designed to measure blood pressure (BP) and activity in freely moving laboratory mice. The fluid filled catheter is placed in the free flowing blood of the systemic artery (inserted into the left carotid artery and extended into the aorta), and the transmitter body is placed in a benign location for long-term biocompatibility. The transmitter can be used to monitor BP in mice (as small as 17 g) under normal physiological and unrestricted conditions 24 h a day while remaining free from stress associated with human interaction. Thus, telemetry is considered the gold standard for BP monitoring in small animals such as mice. However, this methodology does require a good understanding of the system as well as appropriate training to perform the delicate transmitter implantation surgery.

Key words Blood pressure, Radio telemetry, Catheter, Transmitter, Left carotid artery

1 Introduction

A direct intra-arterial assessment of BP in freely moving small laboratory animals such as mice is generally considered the most translationally relevant and the gold standard [1–4]. With the implantable radio telemetry, mice are unrestrained and can be freely moving in their own environment while their BP and activity are being remotely monitored 24 h a day. It is by far the only method which allows the BP to be continuously monitored while remaining free from stress associated with human interaction [5]. Also a diurnal pattern of BP is usually present and can be analyzed in normal mice housed in regular day and light cycle. This is particularly useful if the oscillation of the BP is studied [6]. The method is feasible for measuring blood pressure in both lean and obese mice [7]. However, carrying a transmitter inside the body may negatively influence the development of obesity in mice in a chronic setting (over 6 weeks). Benefit from its high accuracy and low variability, less animals are usually needed for a given statistical

Sean E. Thatcher (ed.), *The Renin-Angiotensin-Aldosterone System: Methods and Protocols*, Methods in Molecular Biology, vol. 1614, DOI 10.1007/978-1-4939-7030-8_7, © Springer Science+Business Media LLC 2017

power as compared to other BP monitoring methods (e.g., tail-cuff method). However, this radio telemetry method requires a delicate surgical procedure for catheter insertion and transmitter implantation. Within this protocol, we demonstrate how to perform transmitter implantation and measure BP using a radio telemetry system in mice.

2 Materials

DSI radio telemetry blood pressure monitoring system for data acquisition and analysis (including platform receivers, matrix and linked computer with installed software), DSI TA11PA-C10 transmitter, equipment and tools for surgical implantation are listed in surgical preparation section.

3 Methods

3.1 Device Operation

1. On and off switch.

   The implantable transmitter has two operational modes and can be switched "on" and "off" by passing a strong magnet. Place a working FM radio in close proximity while turning on the transmitter using a strong magnet. An audible tone from the FM radio indicates a successful "on" operation (*see* **Note 1**).

2. Offset determination.

   Activate the transmitter well advance of the implantation surgery and determine the pressure offset value before placing the device into the animal. Place the transmitter on an empty telemetry receiver, and use the Acquisition system software to determine the BP reading. Normal values are expected to fall between ±3 mmHg. Record the BP offset value to a designated file for later references. If the offset values exceed ±3 mmHg range, send the transmitter back to DSI for exchange or refurbishment.

3. Calibration and configuration

   Each individual transmitter unit has a unique set of calibration values that must be manually configured into the data acquisition system to allow the computer system interpreting signals received from the telemetry device. Those calibration values are associated with the transmitter serial number. Thus, it is important to record the serial number of the transmitter that each mouse received during implantation, and display the number on animal's cage card after surgery (*see* **Note 2**).

3.2 Software Operation (Acquisition and Analysis)

1. Acquisition (this is for offset determination as well as data collection in real experimental setting).

Open Acquisition software, click {Configuration} >> {Hardware}, select an empty receiver. Right click to add {new transmitter}, enter all required information of the transmitter (type, name, serial number, and BP calibration data set). Close configuration wizard and save changes, the newly added transmitter should appear in the main window. Select all target transmitters and right click, select {start sampling} >> {trace and save}. Now the computer system is receiving signals from the transmitter. Record the start and stop date and time. For offset determination, stop sampling after 1 h by selecting all target transmitters and right click, select {stop sampling}. For experimental data collection, stop sampling accordingly (e.g., We typically collect data for 5 consecutive days for each study endpoint).

2. Analysis (this is for data extraction).

   Open Analysis software, click {File} >> {Load data}, click select "Parameters." Select target transmitters from {Form list}. Select time or duration accordingly (refer to the records of starting and stopping time) for data extraction. For offset determination, a time period of half an hour before stop sampling is chosen for data extraction. Record BP values at Mean ± SEM. For experimental data extraction, pulse pressure, systolic BP, mean arterial pressure, diastolic BP, heart rate, and physical activity data can be extracted separately. Also a circadian pattern should be existed in all of these parameters if the animals were housed in normal light dark circle.

3.3 Surgical Preparation

1. As with any small animal survival surgery, sterile instruments and strict aseptic procedure are recommended. Any individual anesthetic protocol should be tested for each animal strain and variation before tempting survival surgery (*see* **Note 3**).
2. Equipment and tools:

   Sterile dissecting microscope.

   Heating pad.

   Isoflurane machine for inhalation anesthesia (Induction 3–5% in O_2, maintenance 1.5–2% in O_2).

   Small needle holder.

   Small, blunt-tipped dissecting scissors.

   Sterile cotton swabs.

   6-0 braided silk suture for skin closure.

   25-gauge syringe needles.

   Curved thumb forceps.

   Vessel dilation forceps (curved and straight, fine tips and smooth finish).

Vessel cannulation forceps (for proper manipulating catheter for insertion into the vessel without causing damage to the catheter).

Veterinary tissue adhesive.

Lint-free gauze pads.

7-0 braided silk for aorta occlusion.

Magnet (switch the transmitter on and off).

Working FM radio.

3.4 Implantation Procedure

1. Flood the catheter channel with sterile saline at room temperature prior to the surgery to dimensionally stabilize the catheter (*see* **Note 4**).
2. After the animals are properly anesthetized, place mouse on a heating pad (*see* **Note 5**), shave and clean the throat area, secure front legs with tape, and allow great access to the vessels along the trachea (Fig. 1).
3. To expose the submandibular throat area, using a small dissecting scissors to make a 3 cm long incision through the skin along the midline (from Lower mandible posteriorly to the sternum). Using two moist sterile surgical swabs gently retract glands and muscles along the tracheal.
4. The left carotid artery can typically be found next to the tracheal and just beneath a thin layer of muscle (Fig. 2). Since the vagus nerve is usually in close proximity to the left carotid artery, great care should be taken to avoid injuring this vital nerve.

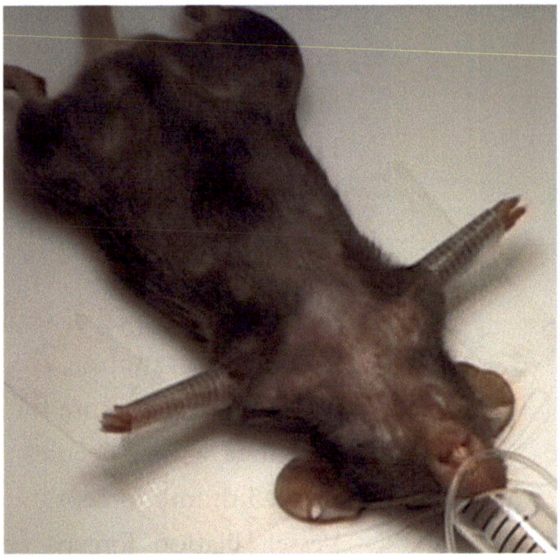

Fig. 1 Proper positioning of the mouse

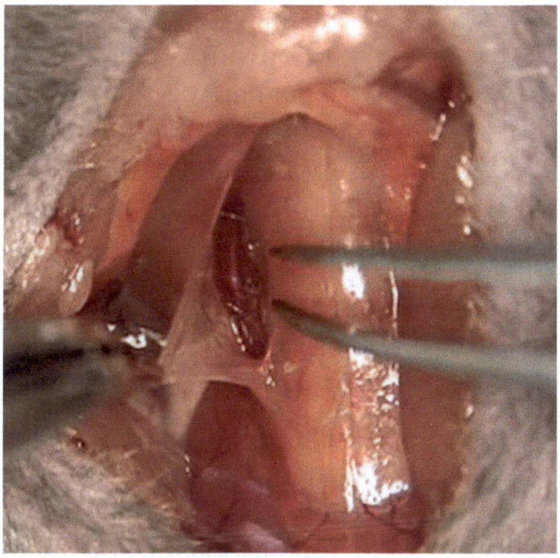

Fig. 2 Isolating left carotid artery

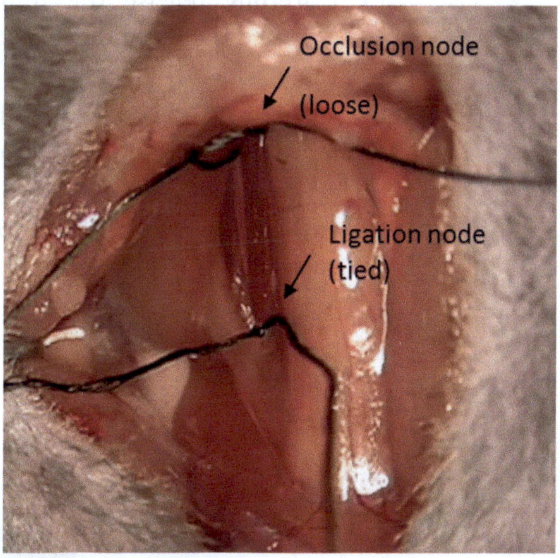

Fig. 3 Ligation (anterior end) and occlusion (posterior end) of the left carotid artery with sutures

5. Carefully isolate the left carotid artery using two vessel dilation forceps and then thread two pieces of 7-0 silk suture beneath the vessel.
6. The more anterior suture will be used to permanently ligate the artery. It is ideally placed just posterior to the bifurcation of the interior and exterior carotid arteries. (This anatomic feature is a useful landmark for determining the proper insertion length for

the implanted catheter). Locate the bifurcation and ligate the artery with a secure surgical node. The suture at the more posterior end will be used temporarily to occlude blood flow in the vessel and allow insertion of the catheter (Fig. 3, *see* **Note 6**).

7. Cannulation of the carotid artery. Bend the tip of a small needle (25 gauge), and use it to insert the catheter (Fig. 4).
8. Remove the protective tip cover from the end of the catheter (if it is new out of box). Holding the non-compliant stem section with a catheter forceps, using thumb forceps grasp the tip cover and remove the tip cover gently. Avoid contact of the catheter to any other surface.
9. Elevate the vessel by applying gentle tension to the ligation and occlusion sutures.
10. Grasp the end of the catheter at the overlap section with vessel cannulation forceps, use the bend-tip needle to insert the catheter, puncture the vessel near the ligation suture, and slip the catheter tip into the carotid artery (Fig. 5, *see* **Note 7**).
11. Carefully release the tension on the occlusion suture and advance the catheter further into the vessel (Fig. 6).
12. Once the catheter is in an ideal position, tie the occlusion suture on the vessel and allow it to seal around the catheter.
13. Take the loose ends of the ligation suture and securely tie them around the notch of the catheter (Fig. 7); this will prevent the catheter from moving once the animals resume normal movement.

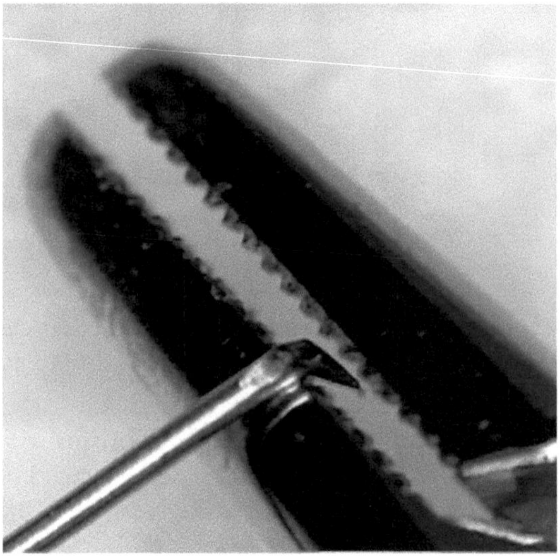

Fig. 4 Bending the needle tip to roughly 80–90°

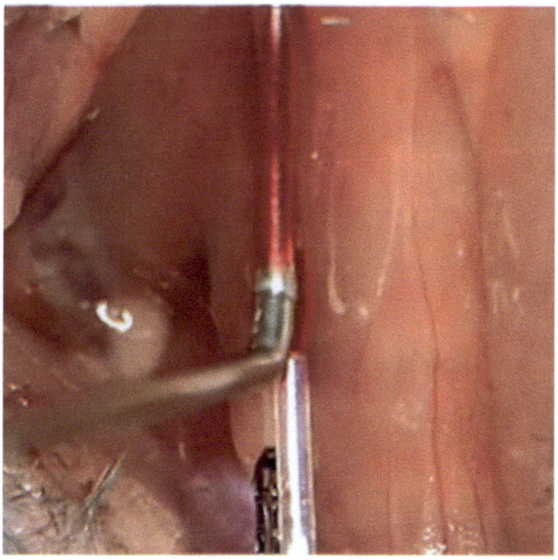

Fig. 5 Puncture the vessel with the bended needed near the ligation suture and slip the catheter tip into the carotid artery

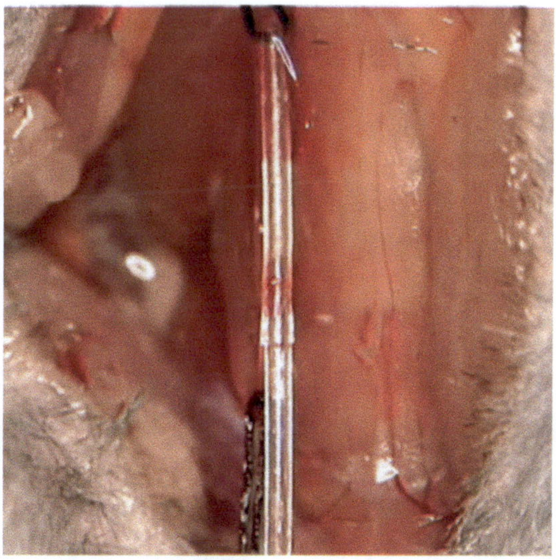

Fig. 6 Slowly loosen the occlusion suture and advance the catheter further into the artery

14. Irrigate the incision area with sterile saline and release the surgical retraction (Fig. 8).
15. Make a pouch underneath the skin along the mouse flank using sterile hemostatic forceps to place the transmitter body (*see* **Note 8**).
16. Once the pouch is formed, irrigate the tunnel using sterile saline and slid in the transmitter (*see* **Note 8**). Use a small drop

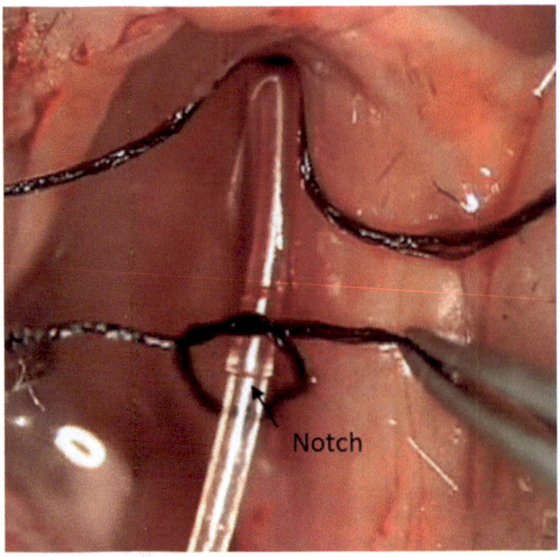

Fig. 7 Secure the catheter in place. Tie the loose occlusion suture on the vessel and seal around the catheter. Tie the loose end of the ligation suture around the notch of the catheter

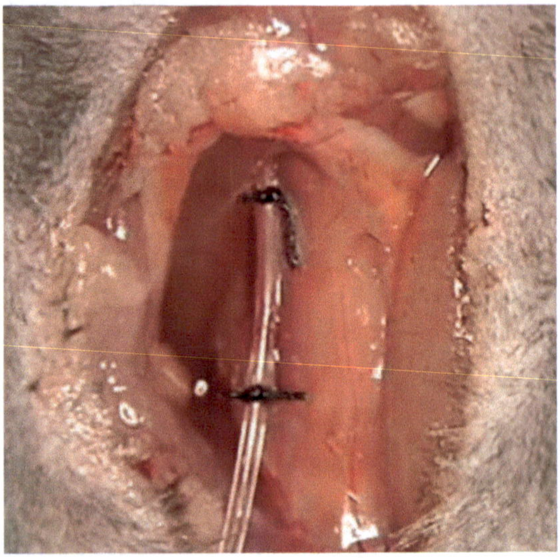

Fig. 8 Clean the loose end of the sutures and irrigate the surgical area

of tissue adhesive to secure the transmitter in place until the connective tissue has a chance to form.

17. To close the skin, irrigate the opening with sterile saline and pull together the tissue edges with 6-0 silk suture and then seal the incision with tissue adhesive.

3.5 Post-Surgical Care

Place the animals back to their own cages on a heating pad after transmitter implantation. Monitor their recover for an hour or two before transfer back to housing room. Check on mice every day and fill out surgery records accordingly. Administer analgesic to surgical mice for the first couple days following implantation according to individual anesthetic protocol. Here, we administer Flunixin (2.5 mg/kg) via subcutaneous injection every 12 h for 2 day following implantation. Typically, for mice that weigh 25 g or more, expect 7–10 days for the mice to resume normal circadian blood pressure rhythm and return to their pre-implantation body weight.

4 Notes

1. Transmitters are shipped in the off mode; only turn on the transmitter prior to its use. Transmitters should be in off mode after each use to save battery life. When switching on or off by passing over a strong magnet, hold both the transmitter and magnet tightly to avoid direct impact.
2. The BP calibration values for each individual transmitter appear on the backside of the sterile case. Record those values together with transmitter serial numbers. Loss of such information will prohibit the use of the transmitter.
3. We usually use isoflurane inhalation as our anesthesia strategy for mice. Different anesthetic methods can be used for this surgical procedure as well. However, each anesthetic protocol should be tested prior to experimentation since it will have a substantial impact on BP recording. Appropriate doses should be given to maintain unconsciousness throughout the whole surgical process (takes 30–45 min per mouse). Over dosing of anesthesia may have negative effects on body temperature; results could delay recovery or even induce death.
4. The catheter (especially the tip area) needs to be clean and straight for a smooth insertion into the vessel. Usually, this is not a problem for new transmitters, but be careful if using a reused transmitter. For reused transmitters, we carefully clean the transmitter (e.g., soaked in Tergazyme enzyme-active powdered detergent to digest any tissue attached to the catheter) under a dissecting microscope with vessel dilation and cannulation forceps without force. The transmitters are then soaked in CIDEX OPA solution for disinfection followed by rinsing with ultrapure water. Thread the whole length of catheter into thin plastic tubing (1 mm in diameter, 6–7 cm in length) and incubate in 37 °C overnight to straighten the catheter.
5. According to our experience, properly maintaining the animal's body temperature has been found to be critical for increasing survival and improving post-surgical recovery.

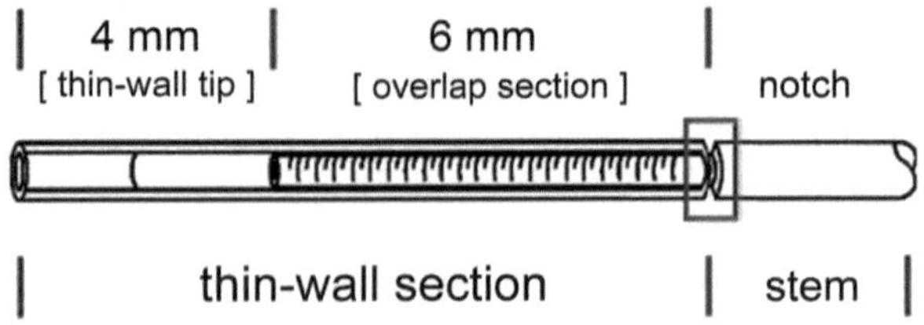

Fig. 9 A close demonstration of the sections of the catheter.

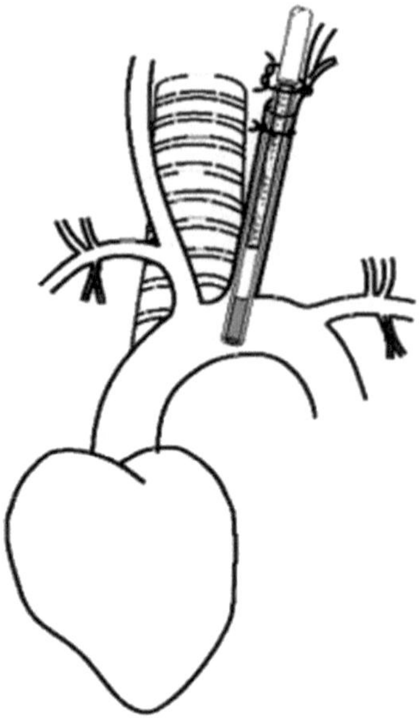

Fig. 10 Position of the tip of the catheter into the free-flowing blood of the aorta

6. The terminal 10 mm of the catheter is composed a thin-wall section of tubing. At the transition point there is a small notch in the profile of the catheter, this notch will be a useful landmark for optimal positioning of the catheter. The tip section of the tubing (treated with anti-thrombogenic film) is designed to be inserted to the animal's blood stream (Fig. 9). It is the distal 4 mm of the catheter that actually transmits the animal's BP. Therefore, it is essential that a significant portion of this thin-wall tip section is placed in the free flow blood of the aorta. Ideally at least 2 mm of the catheter tip should be extended out of the left carotid artery into the free-flowing blood of the aorta. In most mouse models, this can be accomplished by advancing the catheter into the carotid artery until the catheter notch reaches the level of carotid bifurcation (Fig. 10). Minor adjustment may be required based on the size of the mouse. Also, the left carotid artery is chosen to avoid disruption of blood flow to the right branchial subclavian artery which branches off the right carotid artery.

7. The occlusion suture needs to be semi-tied and elevated at the time of puncture and catheter insertion to occlude blood flow. Failure to do so will cause blood to flow out from the puncture site and cause animal death due to excess blood loss. It is normal if only a small amount of blood comes out from the insertion site. Use sterile cotton swabs to clean the blood to retain a clear view.
8. Normally, the transmitter will be placed on the opposite side of the body from the catheter; however for animals that were expected to grow to a size greater than 30 g, place the transmitter on the same side of body as the catheter insertion to reduce stress applied to the catheter as the animal moves about in its cage.

References

1. Brockway BP, Mills P, Kramer K (1998) Fully implanted radio-telemetry for monitoring laboratory animals. Lab Anim 27:40–45
2. Mattson DL (1998) Long-term measurement of arterial blood pressure in conscious mice. Am J Physiol 274(2 Pt 2):R564–R570
3. Van Vliet BN, Chafe LL, Antie V et al (2000) Direct and indirect methods used to study arterial blood pressure. J Pharmacol Toxicol Methods 44(2):361–373
4. Huetteman DA, Boqie H (2009) Direct blood pressure monitoring in laboratory rodents via implantable radio telemetry. Methods Mol Biol 573:57–73
5. Whitesall SE, Hoff JB, Vollmer AP et al (2004) Comparison of simultaneous measurement of mouse systolic arterial blood pressure by radiotelemetry and tail-cuff methods. Am J Physiol Heart Circ Physiol 286(6):H2408–H2415. Epub 2004 February 12
6. Xie Z, Su W et al (2015 Jan) Smooth-muscle BMAL1 participates in blood pressure circadian rhythm regulation. J Clin Invest 125(1):324–336
7. Gupte M et al (2012) Angiotensin converting enzyme 2 contributes to sex differences in the development of obesity hypertension in C57BL/6 mice. Arterioscler Thromb Vasc Biol 32(6):1392–1399

Chapter 8

Characterization and Functional Phenotyping of Renal Immune Cells via Flow Cytometry

Nathan P. Rudemiller and Steven D. Crowley

Abstract

A variety of immune cell subsets contribute to the pathogenesis of hypertension and associated kidney damage following inappropriate activation of the renin–angiotensin system (RAS). These immune cell subsets often express common surface markers, which complicates their separation and characterization in vivo. Accordingly, flow cytometry has become an invaluable tool for parsing immune cell populations because this technique permits the simultaneous detection of up to 18 markers on a single cell. Below we describe a process by which one can determine the immune cell subsets in the kidney via flow cytometry.

Key words Kidney diseases, Inflammation, Flow cytometry, Hypertension, Cytokines

1 Introduction

Dysregulation of the renin–angiotensin system (RAS) contributes to the pathogenesis of hypertension and chronic kidney disease, as evidenced by the efficacy of RAS inhibitors in lowering blood pressure and slowing the progression of kidney diseases in a wide range of patients [1, 2]. Moreover the immune system has been implicated in RAS-mediated renal pathologies [3, 4]. Experimentally both myeloid and lymphoid cells infiltrate the kidney in models of RAS-dependent hypertension and kidney disease and mediate sodium retention, tissue injury, and fibrosis [3, 5, 6]. In order to understand the inflammatory mechanisms that drive kidney dysfunction and injury, it is imperative to characterize and study the specific immune cell subsets that accumulate in the kidney.

Myeloid and lymphoid cells are divided into many functionally distinct subsets. For example, $CD4^+$ T lymphocytes are further divided into four subsets with disparate inflammatory capacities: Th1, Th2, Th17, and T regulatory cells (Treg). Unfortunately

The original version of this chapter was revised. The erratum to this chapter is available at: DOI 10.1007/978-1-4939-7030-8_16

Sean E. Thatcher (ed.), *The Renin-Angiotensin-Aldosterone System: Methods and Protocols*, Methods in Molecular Biology, vol. 1614, DOI 10.1007/978-1-4939-7030-8_8, © Springer Science+Business Media LLC 2017

these subsets cannot be differentiated by single surface markers, but they can be distinguished by multiple surface markers in combination with unique expression profiles of cytokines and/or transcription factors [7–10]. Available immunohistochemical techniques can illuminate the distribution of immune cells in the kidney, but fail to provide the granularity necessary to distinguish specific myeloid and lymphoid subsets. In contrast, flow cytometry is a laser-based technique that permits the simultaneous staining and analysis of up to 18 extracellular or intracellular epitopes on a single cell [11], allowing intricate parsing of immune cell subsets based on phenotypic and functional characteristics. Furthermore fluorescent cell sorting can be employed to separate and collect highly purified cell populations for protein or RNA analysis. Flow cytometry requires that the starting biological sample be a single cell suspension of live cells. This can prove difficult if the cells of interest (e.g., immune cells) are integrated into a solid organ such as the kidney. Therefore, below we outline a methodology to dissociate kidneys—drawing from our experience and the experience of others in the literature—in such a way that will enable detailed analysis of intra-renal immune cells via flow cytometry to further develop our understanding of the immune mechanisms that regulate RAS-dependent renal disease.

2 Materials

Confirm that all tools, media, and work surfaces are sterile.

1. Surgical instruments: Forceps, standard surgical scissors, spring-action micro scissors.
2. Animal clippers.
3. Cotton tip applicators.
4. Gauze.
5. 25 G 5/8 needle.
6. 3 ml syringe.
7. 1 ml syringe plunger.
8. Petri dish (60 × 15 mm).
9. 100, 70, and 40 μm cell strainers.
10. 15 ml conical tubes.
11. Centrifuge with 15 ml conical tube adaptor.
12. Phosphate buffered solution (PBS; 137 mM NaCl, 2.7 mM KCl, 10 mM Na_2HPO_4, 1.8 mM KH_2PO_4; pH 7.4).
13. Deoxyribonuclease I (DNase): reconstitute in PBS at 1 mg/ml, aliquot, and store at −20 °C.
14. Collagenase type IV: reconstitute in PBS at 100 mg/ml, aliquot, and store at −20 °C (*see* **Note 1**).

15. RPMI-10: RPMI 1640 containing 10% fetal bovine serum (FBS).
16. Digestion solution: RPMI 1640 containing 3% FBS, 10 μg/ml DNase, and 1 mg/ml collagenase type IV. Allow digestion buffer to reach room temperature (RT). Add DNase and collagenase immediately before use.
17. Incubator with agitator (37 °C).
18. Wash buffer: PBS containing 2% FBS and 2 mM EDTA. Store at 4 °C.
19. Red blood cell (RBC) lysis buffer (155 mM NH_4Cl, 12 mM $NaCHO_3$, 0.1 mM EDTA).
20. Trypan blue solution, 0.4%.
21. Appropriate fluorochrome-conjugated primary antibodies to stain the cells of interest.
22. Access to a flow cytometer and data analysis software such as FACSDiva or FlowJo.

Optional materials if intracellular staining of inducible factors (e.g., cytokines) is required:

23. (Optional) Atmosphere-controlled incubator (37 °C with 5% CO_2).
24. (Optional) Monensin: reconstitute in DMSO at 10 mg/ml.
25. (Optional) Brefeldin A: reconstitute in DMSO at 10 mg/ml.
26. (Optional) Cell dissociation media (trypsin, EDTA, etc.).

3 Methods

3.1 Collection of Kidney Samples

1. Deeply anesthetize the mouse and shave the abdomen (*see* **Note 2**).
2. Place the anesthetized mouse supine on a surgical table and secure the limbs, thus immobilizing the mouse.
3. Make a longitudinal incision through the skin and abdominal wall spanning from the lower abdomen to the sternum, making sure not to puncture the diaphragm. Make two transverse incisions, bisecting either side of the abdominal wall.
4. Using sterile cotton-tip applicators, gently shift the stomach, intestines, liver, spleen, and any excess fascia laterally to expose the left kidney and vena cava. Upon lifting the sternum with forceps, the heart should be visible through the intact diaphragm.
5. Clasp the sternum with forceps, cut the vena cava below the left renal vein with micro scissors (*see* **Note 3**) and pierce the left ventricle through the intact diaphragm with a 25 G needle attached to a 3 ml syringe filled with ice-cold PBS (*see* **Note 4**).

6. Infuse the PBS through the left ventricle at an approximate rate of 200 μl/s until all 3 ml have been exhausted; absorb the fluid exiting from the lacerated vena cava with gauze. The left kidney can provide a proxy of the efficacy of the PBS flush. The kidney should become pale, indicating that it has been cleared of blood (*see* **Note 5**). Flushing the kidneys ensures that only those immune cells that have infiltrated the renal parenchyma will be included in downstream analysis.
7. Excise and decapsulate the left kidney (*see* **Note 6**). Place the kidney in RPMI-10 on ice. Collect all subsequent kidney samples according to **steps 1–7** (*see* **Note 7**).

3.2 Mechanical and Enzymatic Digestion of Kidneys

1. Place a 100 μm cell strainer in a petri dish. Aliquot 5 ml of RT digestion buffer into the filter.
2. Place a kidney in the filter and mince it with a pair of scissors into several small pieces. With the blunt end of a 1 ml syringe plunger, grind the kidney pieces into the filter until the tissue is homogenized.
3. Overturn the filter and hold it over the petri dish. Scoop the kidney homogenate into the petri dish with the syringe plunger. Rinse the overturned filter with 1–2 ml digestion buffer while holding it over the petri dish to collect residual sample (*see* **Note 8**).
4. Mix the contents of the petri dish and transfer the contents to a 15 ml conical tube using a transfer pipette. Incubate the kidney homogenate under constant agitation for 30 min at 37 °C.
5. Remove the sample from 37 °C and place it on ice. Add 3 ml ice-cold wash buffer to the 15 ml conical tube containing the sample and invert the tube several times to mix the solution.
6. Centrifuge the sample for 1 min at 10 × *g* to pellet heavy debris.
7. Rinse a 70 μm cell strainer with wash buffer. Pour the kidney homogenate through the rinsed filter into a new 15 ml conical tube and discard the pellet of heavy debris. Centrifuge the sample for 7 min at 300 × *g*.
8. Discard the supernatant and suspend the cell pellet in 3 ml RBC lysis buffer to eliminate residual RBCs. Incubate cells for 5 min on ice. Add 5 ml wash buffer and invert the tube to mix.
9. Rinse a 40 μm cell strainer with wash buffer. Pour the sample through the cell strainer into a new 15 ml conical tube. Centrifuge the cells for 7 min at 300 × *g* (*see* **Note 9**).
10. Suspend the cell pellet in 5 ml wash buffer. Add 10 μl of the cell suspension to a microcentrifuge tube containing 10 μl Trypan blue solution and pipette to mix. Determine the cell concentration by counting live (Trypan blue-negative) cells on a hemocytometer (*see* **Notes 10** and **11**), and use this concen-

tration to determine the total number of live cells in your sample—a value that can be utilized in the data analysis to quantify immune cell subsets.

3.3 (Optional) Activation of Renal Immune Cells to Elicit Measureable Cytokine Production and Other Inducible Factors for Antibody Staining (See Note 12)

1. Wash cells once with PBS to remove EDTA that may carry over from the wash buffer. Suspend cells at 2×10^6 cells/ml in RPMI-10 containing the protein transport inhibitors monensin and brefeldin A (5 μg/ml each) (*see* **Note 13**). Aliquot cells into a 24-well culture dish at 1 ml/well and add an appropriate cell stimulant. For example, the combination of PMA (10 ng/ml) and ionomycin (1 μg/ml) has been used to induce the production of TNF, IFNG, and IL-17a in renal T cells [12–14], while LPS can be used to stimulate iNOS production in renal monocytes/macrophages [6]. Prepare duplicate wells for each sample as well as control wells that receive protein transport inhibitor but no stimulation. Place the samples in a cell culture incubator for 4–6 h (*see* **Note 14**).
2. Following incubation, combine the desired cells from the duplicate wells into 12 × 75 mm round-bottom test tubes. For suspension cells, such as T cells, transfer the supernatant. To collect adherent cells, such as myeloid cells, aspirate the supernatant, wash the wells with PBS, and remove the adherent cells from the wells with cell dissociation media.

3.4 Staining Cell Surface Markers and/or Intracellular Antigens for Flow Cytometry

Once the kidney has been dissociated into a single-cell suspension, the method for staining the cells for flow cytometry is no different from staining any other single-cell population. This methodology, utilizing fluorochrome-conjugated antibodies that recognize extracellular or intracellular epitopes, has been thoroughly described elsewhere, and may vary slightly depending on the antibody manufacturer. We recommend consulting the manufacturer website for the appropriate staining protocol. Therefore, instead of describing the staining procedure, below we suggest an epitope panel to measure the capacity of renal T cells to produce TNF and IFNG (Table 1). This epitope panel requires that the flow cytometer is equipped with a violet laser (roughly 405 nm), a blue laser (roughly 488 nm), a red laser (roughly 633 nm), and all of the appropriate filters to detect the listed fluorochromes. The number and types of fluorochromes that one can use will depend on the spectral overlap of the fluorochromes and the cytometer's capability to excite the fluorochromes and detect their emittances. Consult with the personnel of the flow cytometry core regarding the equipment capability and read further about the construction of an antibody panel [15–18]. Additionally fluorescence-minus-one (FMO) controls should be prepared for each antibody in the panel. FMO controls are experimental cells that have been stained with all but one of the fluorochromes in the panel. These controls are essential

Table 1
Seven-color antibody panel to measure T cell cytokine production

Fluorochrome conjugate	Primary antibody (clone)
BV421	CD4 (RM4-5)
BV510	CD45 (30-F11)
FITC	CD3ε (145-2C11)
PE	IFNG (XMG1.2)
PE-cy7	CD8α (53-6.7)
APC	TNF (MP6-XT22)
Near IR dead cell stain	–

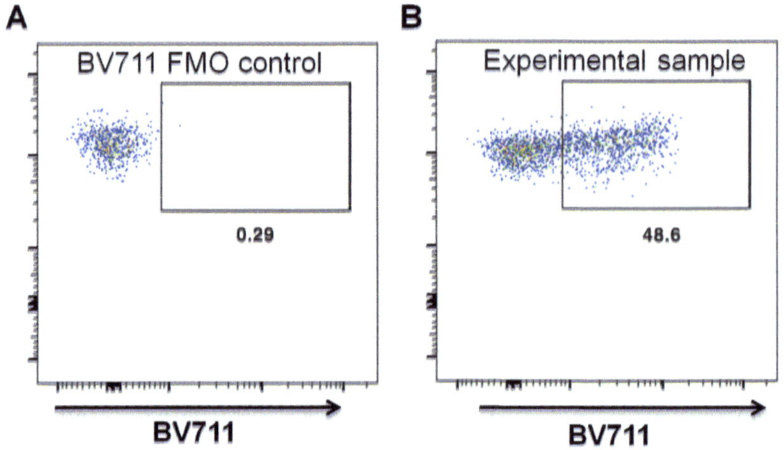

Fig. 1 Using FMO controls to determine positive staining. The FMO control (**a**) is stained with all but one fluorochrome-conjugated antibody in the staining panel. In this example, the FMO for the antibody conjugated to BV711 allows the appropriate determination of positive staining in experimental samples (**b**) with no clear distinction between the positive and negative populations

in determining positive staining, especially for sweeping epitope expression without a clear distinction between the negative and positive populations (Fig. 1). A completely unstained sample will not suffice in determining positive staining since the presence of other fluorochromes in the FMO may shift the negative population due to emission spillover.

3.5 Analysis of Flow Cytometry Data from Renal Immune Cells

The purpose of this section is not to explain how to acquire data on a flow cytometer, as that process will vary depending on the cytometer and software used to collect the data and has been explained elsewhere [16, 18]. We would, however, like to suggest a gating strategy to analyze the cytokine expression of renal T cells from data generated using the epitope panel in Table 1. This gating strategy is illustrated in Figs. 2 and 3. Gates 1A-1D will be common among most staining panels of renal immune cells, not just the panel presented in this protocol. The subsequent gates will vary depending upon the additional antibodies in the panel. The data in Fig. 2 were generated with renal cells from a 129/SvEv mouse after 14 days of chronic angiotensin II infusion, collected on a BD LSRII cytometer, and analyzed using FlowJo software. Each dot on the scatter plots represents a recorded event. These events are not necessarily cells, as the cytometer does not discriminate between debris, cells, etc. Therefore, the data will be referred to as events until we reach a point in the analysis where debris has been excluded.

1. Place side scatter area (SSC-A), a measure of granularity or internal complexity, on the *Y*-axis and near-IR dead cell stain on the *X*-axis to gate the events that do not have positive fluorescence (Fig. 2a). The intensity of the fluorescence of the

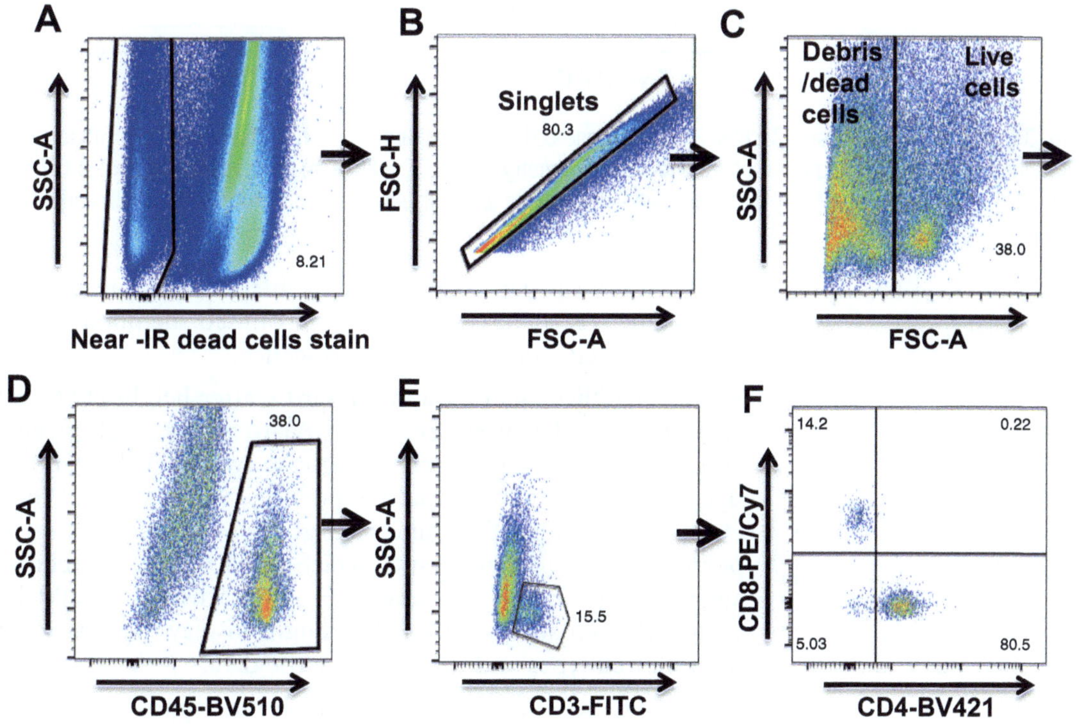

Fig. 2 Gating strategy to analyze renal T cells. All events collected on the flow cytometer are first assessed for expression of dead cell stain (**a**), and those events with positive expression are excluded from downstream analysis. Live events are further narrowed to single events (**b**). Debris and residual dead cells are excluded within the single events (**c**). Live cells are then selected for expression of CD45 (**d**), followed by CD3 (**e**), and then parsed by expression of CD4 and CD8 (**f**)

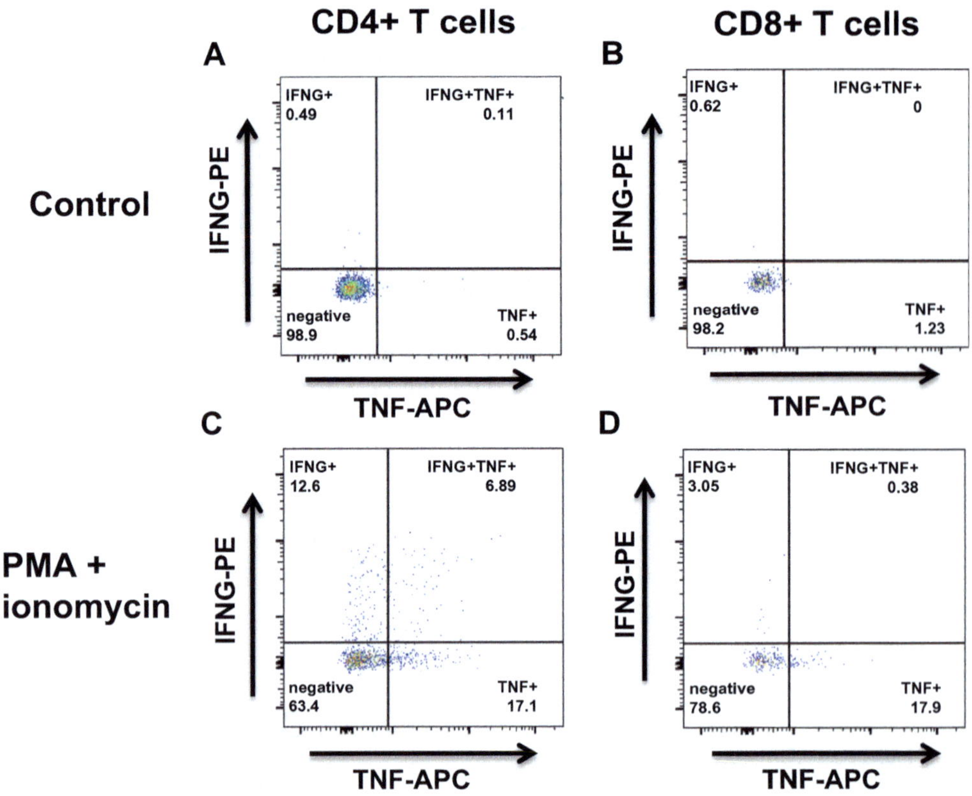

Fig. 3 Cytokine profile of renal T cells. Renal immune cells were cultured for 6 h in the presence or absence of PMA/ionomycin and subsequently stained for flow cytometry. Control T cells (**a**, **b**) showed little expression of TNF and IFNG whereas T cells stimulated with PMA + ionomycin (**c**, **d**) displayed robust expression of TNF and IFNG

dead cell stain is dependent on its ability to permeate the cell membrane. As such, cells with compromised membranes will (e.g., dead cells) display high fluorescence.

2. Within the "live" gate, graph forward scatter height (FSC-H) against forward scatter area (FSC-A) to select single events (Fig. 2b). These two parameters indicate event signal and event size, respectively, and should report the same value. When events fall outside of the linear range, this indicates a disproportionate size to signal ratio, which often represents cellular aggregates (doublets, triplets, etc.). Those events should be excluded from analysis since they might display markers from multiple cell types and thus obscure the data.
3. In order to set the singlet gate, which may appear unclear at first, one can utilize "back-gating" (Shown in Fig. 4). By gating a large, labeled cell population, such as $CD45^+$ events, and displaying that population in the FSC-H vs. FSC-A plot (Fig. 4a), one can clearly set a singlet gate that can then be applied to the upstream population, e.g., live events (Fig. 4b).

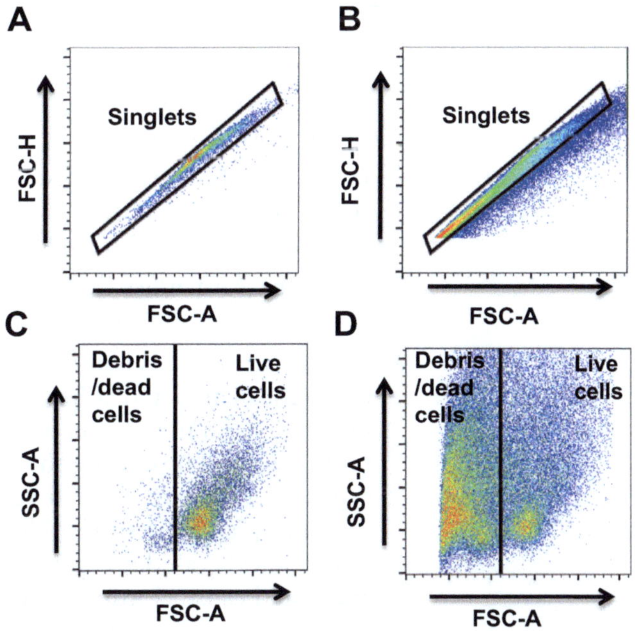

Fig. 4 Performing back-gating. CD45+ events were gated for single events (**a**), which was then be applied to the upstream "live event" population (**b**). Likewise, CD45+ events were gated to exclude debris/dead cells (**c**), and this gate was applied to the upstream singlet population (**d**)

4. Within the singlet gate, graph SSC-A against FSC-A to exclude debris and residual dead cells (Fig. 2c) (*see* **Note 15**). Debris/dead cells can be discriminated by their low FSC-A. Again this gate can be determined by back-gating using the $CD45^{+}$ population (Fig. 4c and d).
5. Within the live cell gate, graph SSC-A against CD45-BV510 to select all $CD45^{+}$ cells, or leukocytes (Fig. 2d).
6. Within the $CD45^{+}$ gate, graph SSC-A against CD3-FITC to select $CD3^{+}$ cells, or T cells (Fig. 2e). The FMO control was essential in setting this gate.
7. Within the T cell gate, graph CD8-PE/Cy7 against CD4-BV421 to differentiate CD8+, CD4+, and double negative T cells (Fig. 2f).
8. Within the $CD4^{+}$ gate and $CD8^{+}$ gate, graph TNF-APC against IFNG-PE to determine the percentage of these populations that can produce TNF and/or IFNG (Fig. 3). Both unstimulated (Fig. 3a and b) and stimulated (Fig. 3c and d) samples from the same kidney are shown to illustrate the necessity of a stimulant to elicit the generation of these cytokines for measurement.

4 Notes

1. Collagenase Type IV is comparable to Collagenase D and Collagenase CLS-4.
2. It is essential that the mouse be anesthetized during, not euthanized before, the kidney harvest in order to reduce blood clotting and cell death in the kidney.
3. The vena cava must be punctured to allow fluid to exit the vascular network during the infusion.
4. The left ventricle is sometimes difficult to visualize and pierce through an intact diaphragm. Alternatively incisions can be made through the ribs on both sides of and parallel to the sternum, followed by cutting the diaphragm away, thus completely exposing the heart. The left ventricle can then be accessed with ease and pierced with the 25 G needle. This process must be done quickly, as deoxygenated blood will begin to circulate and could cause clotting issues in peripheral tissue.
5. In some forms of kidney disease, microvascular damage may occur. In this case, the kidney may appear splotchy. This may not be due to inefficient flushing, but the inability to clear blood due to injury in the vascular architecture.
6. The right kidney will be cleared of blood as well and can be utilized in this protocol if the left kidney is required for other forms of analysis (histology, RNA isolation, etc.).
7. Keep in mind that the number of samples collected and the pace of collection may affect cell viability, as the precedent kidney samples will be incubating on ice during all subsequent sample collections. Intersperse the collection of experimental and control kidneys to account for this temporal factor.
8. Alternatively automatic tissue dissociators such as the gentleMACS Dissociator may be used to streamline the dissection process and reduce user variability during kidney dissociaton.
9. Following this step, many groups subject the cells to a percoll gradient, which separates particles based on density [13, 14]. The utility of this gradient is to further enrich for leukocytes by exploiting their known density range. However, in our hands, we are able to record a sufficient number of renal immune cells via flow cytometry without percoll enrichment, and others have reported decreased cell recovery and reduced viability when using density enrichment [19]. Therefore, we suggest to perform this density enrichment only if necessary, for example, to capture a rare cell population.
10. An aliquot of the cell solution can be diluted further if the sample is too concentrated to accurately count.

11. Much debris may be present in your sample. This is not unexpected, though the parsing of cell from debris may be difficult while counting on a hemocytometer. Look for cellular characteristics such as a circular shape or the presence of a nucleus to identify cells.
12. Ex vivo activation of the renal immune cells is necessary only if the measurement of inducible factors, such as cytokines, is required.
13. The addition of protein transport inhibitors is necessary only if secreted factors such as cytokines are being measured via intracellular staining. Furthermore, monenesin and brefedin A have each been shown to effect the expression of particular surface makers or intracellular cytokines [20, 21]. This should be taken into consideration during experimental design.
14. The user must determine the appropriate cell stimulant and time of incubation to elicit the inducible factor of interest from the relevant cell type.
15. The live cell population represented in Fig. 2c is analogous to the initial cell count performed on the hemocytometer at the end of the mechanical digestion protocol. Because live, single cells were counted on the hemocytometer, one can assume that the total count of cells is represented in the live cell gate in Fig. 2c. Therefore, the total number of any following gates can be determined by simple mathematics using the gate percentages afforded by the data analysis program. This type of quantification is appropriate only when performing flow cytometry staining immediately after counting, as manipulations, such as activation in culture, skew the population due to cell adherence to the plate, increased apoptosis, etc.

References

1. Dahlof B, Devereux RB, Kjeldsen SE, Julius S, Beevers G, de Faire U, Fyhrquist F, Ibsen H, Kristiansson K, Lederballe-Pedersen O, Lindholm LH, Nieminen MS, Omvik P, Oparil S, Wedel H, Group LS (2002) Cardiovascular morbidity and mortality in the losartan intervention for endpoint reduction in hypertension study (LIFE): a randomised trial against atenolol. Lancet 359(9311):995–1003. doi:10.1016/S0140-6736(02)08089-3
2. Brenner BM, Cooper ME, de Zeeuw D, Keane WF, Mitch WE, Parving HH, Remuzzi G, Snapinn SM, Zhang Z, Shahinfar S, Investigators RS (2001) Effects of losartan on renal and cardiovascular outcomes in patients with type 2 diabetes and nephropathy. N Engl J Med 345(12):861–869. doi:10.1056/NEJMoa011161
3. Zhang JD, Patel MB, Griffiths R, Dolber PC, Ruiz P, Sparks MA, Stegbauer J, Jin H, Gomez JA, Buckley AF, Lefler WS, Chen D, Crowley SD (2014) Type 1 angiotensin receptors on macrophages ameliorate IL-1 receptor-mediated kidney fibrosis. J Clin Invest 124(5):2198–2203. doi:10.1172/JCI61368
4. Zhang JD, Patel MB, Song YS, Griffiths R, Burchette J, Ruiz P, Sparks MA, Yan M, Howell DN, Gomez JA, Spurney RF, Coffman TM, Crowley SD (2012) A novel role for type 1 angiotensin receptors on T lymphocytes to limit target organ damage in hypertension. Circ Res 110(12):1604–1617. doi:10.1161/CIRCRESAHA.111.261768
5. Guzik TJ, Hoch NE, Brown KA, McCann LA, Rahman A, Dikalov S, Goronzy J, Weyand C, Harrison DG (2007) Role of the T cell in

the genesis of angiotensin II induced hypertension and vascular dysfunction. J Exp Med 204(10):2449–2460. doi:10.1084/jem.20070657

6. Zhang J, Rudemiller NP, Patel MB, Karlovich NS, Wu M, McDonough AA, Griffiths R, Sparks MA, Jeffs AD, Crowley SD (2016) Interleukin-1 receptor activation potentiates salt reabsorption in angiotensin II-induced hypertension via the NKCC2 Co-transporter in the nephron. Cell Metab 23(2):360–368. doi:10.1016/j.cmet.2015.11.013
7. Korn T, Bettelli E, Oukka M, Kuchroo VK (2009) IL-17 and Th17 cells. Annu Rev Immunol 27:485–517. doi:10.1146/annurev.immunol.021908.132710
8. Szabo SJ, Kim ST, Costa GL, Zhang X, Fathman CG, Glimcher LH (2000) A novel transcription factor, T-bet, directs Th1 lineage commitment. Cell 100(6):655–669
9. Hori S, Nomura T, Sakaguchi S (2003) Control of regulatory T cell development by the transcription factor Foxp3. Science 299(5609):1057–1061. doi:10.1126/science.1079490
10. Zheng W, Flavell RA (1997) The transcription factor GATA-3 is necessary and sufficient for Th2 cytokine gene expression in CD4 T cells. Cell 89(4):587–596
11. Chattopadhyay PK, Roederer M (2012) Cytometry: today's technology and tomorrow's horizons. Methods 57(3):251–258. doi:10.1016/j.ymeth.2012.02.009
12. Itani HA, Xiao L, Saleh MA, Wu J, Pilkinton MA, Dale BL, Barbaro NR, Foss JD, Kirabo A, Montaniel KR, Norlander AE, Chen W, Sato R, Navar LG, Mallal SA, Madhur MS, Bernstein KE, Harrison DG (2016) CD70 exacerbates blood pressure elevation and renal damage in response to repeated hypertensive stimuli. Circ Res. doi:10.1161/CIRCRESAHA.115.308111
13. Saleh MA, McMaster WG, Wu J, Norlander AE, Funt SA, Thabet SR, Kirabo A, Xiao L, Chen W, Itani HA, Michell D, Huan T, Zhang Y, Takaki S, Titze J, Levy D, Harrison DG, Madhur MS (2015) Lymphocyte adaptor protein LNK deficiency exacerbates hypertension and end-organ inflammation. J Clin Invest 125(3):1189–1202. doi:10.1172/JCI76327
14. Wei Z, Spizzo I, Diep H, Drummond GR, Widdop RE, Vinh A (2014) Differential phenotypes of tissue-infiltrating T cells during angiotensin II-induced hypertension in mice. PLoS One 9(12):e114895. doi:10.1371/journal.pone.0114895
15. Radcliff G, Jaroszeski MJ (1998) Basics of flow cytometry. Methods Mol Biol 91:1–24
16. Maecker HT, Trotter J (2006) Flow cytometry controls, instrument setup, and the determination of positivity. Cytometry A 69(9):1037–1042. doi:10.1002/cyto.a.20333
17. Shapiro HM (2004) Excitation and emission spectra of common dyes. Curr Protoc Cytom Chapter 1:Unit 1.19. doi:10.1002/0471142956.cy0119s26
18. Roederer M (2001) Spectral compensation for flow cytometry: visualization artifacts, limitations, and caveats. Cytometry 45(3):194–205
19. Yu YR, O'Koren EG, Hotten DF, Kan MJ, Kopin D, Nelson ER, Que L, Gunn MD (2016) A protocol for the comprehensive flow cytometric analysis of immune cells in normal and inflamed murine non-lymphoid tissues. PLoS One 11(3):e0150606. doi:10.1371/journal.pone.0150606
20. Schuerwegh AJ, Stevens WJ, Bridts CH, De Clerck LS (2001) Evaluation of monensin and brefeldin A for flow cytometric determination of interleukin-1 beta, interleukin-6, and tumor necrosis factor-alpha in monocytes. Cytometry 46(3):172–176
21. Nylander S, Kalies I (1999) Brefeldin A, but not monensin, completely blocks CD69 expression on mouse lymphocytes: efficacy of inhibitors of protein secretion in protocols for intracellular cytokine staining by flow cytometry. J Immunol Methods 224(1–2):69–76

Chapter 9

Assessment of the Renin–Angiotensin System in Cellular Organelle: New Arenas for Study in the Mitochondria

Bryan A. Wilson and Mark C. Chappell

Abstract

The renin–angiotensin system (RAS) is an important hormonal system composed of various protein and peptide components that contribute to blood pressure regulation. Although originally characterized as a circulating system, there is increasing evidence for the intracellular expression of RAS elements on the nucleus and mitochondria that may function in concert with or independent of the circulating system. The present chapter describes several experimental approaches to quantify the expression of RAS components in isolated mitochondria from the kidney. These approaches are intended to provide a framework to understand the mitochondrial RAS within a cell-free environment.

Key words Mitochondria, Renin–angiotensin system, Subcellular fractionation, Peptide metabolism, Ang II, Ang-(1–7), Renin, Mas protein, Neprilysin, Thimet oligopeptidase

1 Introduction

The renin–angiotensin system (RAS) is a key hormonal system that regulates blood pressure through both central and peripheral mechanisms [1]. It is also quite clear that the RAS constitutes a multifunctional set of bioactive peptides, processing enzymes, and distinct receptor subtypes [2]. Figure 1 outlines these pathways of the RAS that ultimately begins with conversion of the precursor protein angiotensinogen to Ang I by the aspartyl protease renin and the subsequent conversion to Ang II by the metallo-dipeptidylcarboxpeptidase angiotensin I converting enzyme (ACE). Ang II binds to and activates the AT1 receptor (AT1R) subtype to elicit a number of actions that include an increase in blood pressure, vasoconstriction, tissue fibrosis, cellular proliferation, and inflammation (Fig. 1). Ang II also recognizes the AT2R subtype that generally opposes actions of the AT1R including vasodilation mediated through the release of nitric oxide (NO). Ang II undergoes further processing by the mono-aminopeptidases to form Ang-(2–8) [Ang III] and Ang-(3–8) [Ang IV]. Ang III

Sean E. Thatcher (ed.), *The Renin-Angiotensin-Aldosterone System: Methods and Protocols*, Methods in Molecular Biology, vol. 1614, DOI 10.1007/978-1-4939-7030-8_9, © Springer Science+Business Media LLC 2017

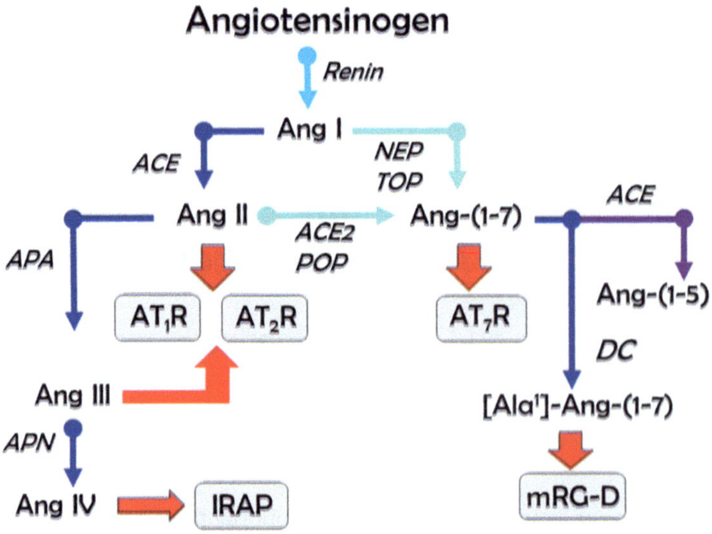

Fig. 1 Processing pathways within the renin-angiotensin system. The protease renin converts angiotensinogen to Ang I. In the classical pathway, ACE converts Ang I to Ang II which binds to both AT1R and AT2R. Ang II is further converted by aminopeptidase A (APA) to Ang III that may interact with AT1R and AT2R. Ang III is converted by aminopeptidase N to Ang-(3–8) [Ang IV] that interacts with the insulin regulated aminopeptidase (IRAP). In the alternative pathway, Ang I is processed to Ang-(1–7) by the endopeptidases neprilysin (NEP) and thimet oligopeptidase (TOP). Ang-(1–7) is also derived from processing of Ang II by ACE2 and prolyl oligopeptidase (POP). Ang-(1–7) recognizes the AT7R/Mas protein. Ang-(1–7) is metabolized by ACE to Ang-(1–5). Ang-(1–7) can be modified by an aspartyl decarboxylase (DC) to form [Ala^1]-Ang-(1–7) that recognizes the Mas-related receptor mRG-D. Adapted from Chappell, American Journal of Physiology—Heart and Circulatory Physiology 310: H137–H152, 2016

recognizes both AT1R and AT2R, while the hexapeptide Ang IV apparently interacts with insulin regulated aminopeptidase (IRAP) (Fig. 1). Ang I is also directly processed to Ang-(1–7) by several endopeptidases including neprilysin (NEP) and thimet oligopeptidase (TOP) through hydrolysis of the Pro7-Phe8 bond; the monocarboxypeptidases ACE2 and prolyl oligopeptidase (POP) generate Ang-(1–7) from Ang II (Fig. 1). The heptapeptide Ang-(1–7) primarily recognizes the AT7R/Mas protein to elicit vasodilation through NO release, as well as attenuate fibrosis, inflammation, proliferation oxidative stress, and epithelia to myofibroblast transition [2, 3]. Ang-(1–7) may undergo further modification through decarboxylation of the aspartyl residue to form Ala1-Ang-(1–7) [almandine] that recognizes the Mas-related receptor mRG-D to induce vasorelaxation (Fig. 1) [3]. The importance of these "alternative pathways" is most likely evident during therapeutic blockade of the RAS as ACE inhibitors markedly increase circulating levels of Ang-(1–7) while AT1R antagonists (ARBs) increase both

Ang II and Ang-(1–7) to stimulate AT2R and AT7R, respectively. Although the activation of the RAS is typically considered to reflect the binding of Ang II to the extracellular AT1R, substantial evidence suggests an intracellular pool of functional receptors localized to the nucleus, endoplasmic reticulum and mitochondria [4–15]. Indeed, the intracellular administration of Ang II to cardiac myocytes, vascular smooth muscle cells and proximal tubule epithelial cells results in an immediate increase in the intracellular levels of calcium [9, 16, 17]. Moreover, Ang II elicits a rapid increase in reactive oxygen species (ROS) that is blocked by the ARBs losartan and candesartan, as well as the NAD(P)H oxidase inhibitor diphenyliodium in nuclei isolated from the rat and sheep renal cortex [6, 10]. The direct application of Ang II to isolated nuclei also elicits an increase in the mRNA levels of angiotensinogen, renin, MCP-1, and NHE-3 [18]. Intracellular angiotensin receptors do not solely comprise the AT1R subtype as both functional AT2R and AT7R/Mas linked to NO stimulation were identified on the nucleus and mitochondria [4, 5, 7, 10, 11]. Abadir and colleagues presented evidence for an intra-mitochondrial RAS by identifying AT1 and AT2 receptors within isolated liver mitochondria of mice; Ang II activation of mitochondrial AT2R stimulated local levels of NO [4, 8]. The expression of the AT1 to AT2 receptor ratio was higher in the mitochondria of aged animals, and the chronic treatment of the aged mice with the ARB losartan mitigated the enhanced expression of the AT1R [19].

We expanded the characterization of an intra-mitochondrial RAS through the demonstration of an Ang-(1–7)-AT7R axis in mitochondria isolated from the sheep renal cortex [7]. Renal mitochondria expressed angiotensinogen, active renin, and the peptides Ang-(1–7), Ang II, and Ang I [29]. Additionally, these studies revealed the processing of Ang I to Ang-(1–7) in isolated mitochondria that reflected the contribution of the endopeptidases NEP and TOP [7].

Lastly, we demonstrated immunoreactive evidence for expression of the AT7R/Mas in cortical mitochondria. These studies collectively support the existence of an Ang-(1–7) pathway within mitochondria and the possibility for the peptide to elicit direct actions on mitochondrial function. Indeed, the beneficial aspects of the Ang-(1–7) axis may counteract the deleterious actions of the Ang II-AT1R pathway within the mitochondria [7].

Due to emerging evidence for the complex interplay between mitochondrial function and the RAS, we review various methodologies to assess the RAS components in isolated mitochondria. The intent of the chapter is to provide a biochemical approach to elaborate the mitochondrial components and functions in a cell-free context. We detail mitochondrial isolation from the kidney, as well as describe methodologies to assess mitochondrial purity, renin activity, and angiotensin peptide content and metabolism.

Mitochondria are isolated utilizing a combination of differential centrifugation and densitometric separation by a discontinuous Percoll gradient. The separation of mitochondria and other organelles on Percoll is a widely utilized method that typically retains the functional integrity of the organelle [4, 7, 20, 21]. Moreover, isolation of mitochondrial on Percoll gradients is rapidly obtained from intact tissues (<2 h) and can be achieved with a high-speed centrifuge obviating the necessity of an ultracentrifuge. We routinely utilize a Percoll density gradient consisting of a 24% and 40% layered suspensions (Fig. 6). The crude mitochondria pellet derived from the differential centrifugation separation is reconstituted in 12% Percoll/buffer A and gently layered on top of the 24% interface. Following centrifugation at 15,000 × g for 5 min, the mitochondria sediments in a visible band between the 24% and 40% Percoll layers. The purity and enrichment of the isolated mitochondria are assessed by immunoblot analysis using antibodies against the following mitochondrial proteins: voltage-dependent anion channel (VDAC), an outer mitochondrial membrane marker, and the inner mitochondrial markers of the electron transport chain: ATP synthase complex V (CV-ATP5A); ubiquinolcytochrome c reductase core protein II complex III (CIII-UCRC2); and succinate dehydrogenase complex subunit B complex II (CII-SDHB) (Fig. 2a–c). Additionally, a cathepsin B antibody is utilized as a marker for lysosomes and renin granules, as well a β-actin antibody for cytosolic contamination [22].

Renin constitutes the rate-limiting enzyme catalyzing the processing of angiotensinogen to Ang I (Fig. 1). Peters and colleagues originally demonstrated an alternative cytosolic transcript of renin that was predicted to be retained within the cell and not enter the secretory pathway [23]. This truncated form of renin was active and preferentially localized to the mitochondria. Interestingly, the renin isoform was actively taken up and internalized by isolated mitochondria, but full-length prorenin did not undergo internalization [23, 24]. Ishigami et al. recently identified a constitutively active form of renin that lacks a portion of the pro-segment in the proximal tubules of the mouse kidney [25]. In transgenic mice, the forced overexpression of this renin isoform increased blood pressure, but did not elevate circulating renin [25].

Based on these earlier studies, we assessed renin activity in isolated mitochondria from male and female sheep by quantifying the disappearance of intact angiotensinogen (Fig. 3). In contrast to typical renin assays that quantify the formation of Ang I, our assay was facilitated by the specificity of the antibody to recognize the N-terminal Ang I sequence of angiotensinogen. As shown in the immunoblot in Fig. 3a–d, renin activity in isolated mitochondria was evident by the time-dependent disappearance of intact angiotensinogen at 37 °C; angiotensinogen processing was abolished by pretreatment of the mitochondria with aliskiren (Fig. 3g and h).

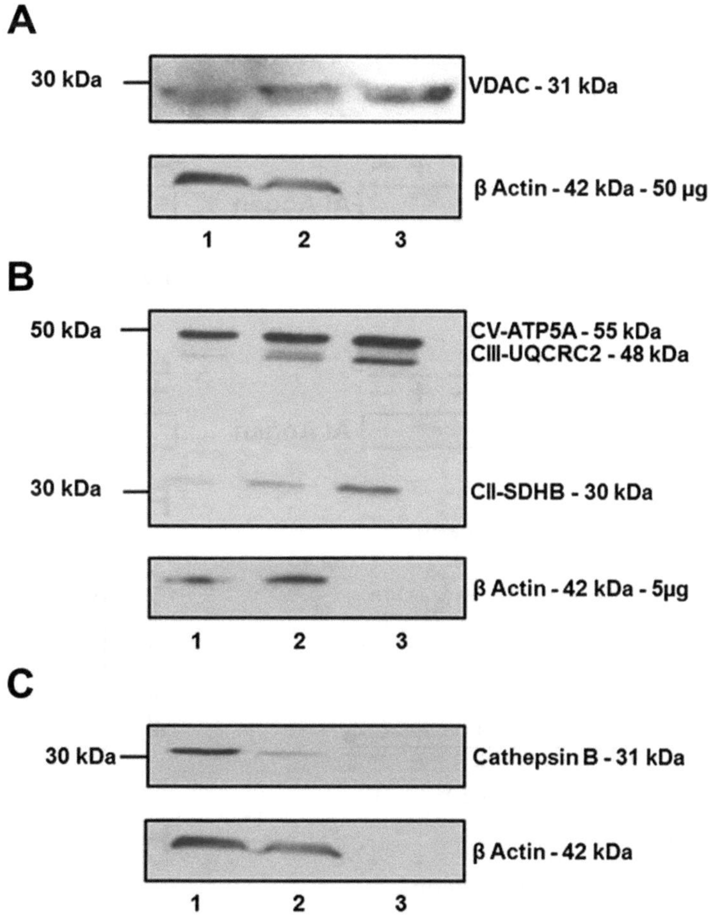

Fig. 2 Assessment of mitochondria purity from isolated fractions of the sheep renal cortex. Whole renal cortex homogenate (*lane 1*), crude mitochondrial fraction from a 10,000 × *g* differential centrifugation step (*lane 2*), and the Percoll-purified mitochondrial fraction (*lane 3*) were subjected to 10% SDS gel and immunoblotting with anti-VDAC (**a**—50 μg total protein) as well as ATP5A, UQCRC2, SDHB (**b**—5 μg total protein) and Cathepsin B (**c**—50 μg total protein), as well as β-actin. Purified mitochondria appear to exhibit greater expression of both the outer mitochondria marker (VDAC) and the inner mitochondrial markers (ATP5A, UQCRC2, and SDHB), while Cathepsin B and β-actin expression were diminished or absent. Data adapted from Wilson et al., *Am J Physiol Renal Physiol* 310: F637–F645, 2016

To determine prorenin levels, mitochondria were pretreated with trypsin to convert prorenin to active renin, and angiotensinogen processing was determined in the presence of the trypsin inhibitor SBTI. Trypsin exposure did not appear to markedly increase renin activity as the disappearance rate of Ang I-angiotensinogen was comparable for both the trypsin and non-trypsin treated mitochondrial samples (Fig. 3c–f).

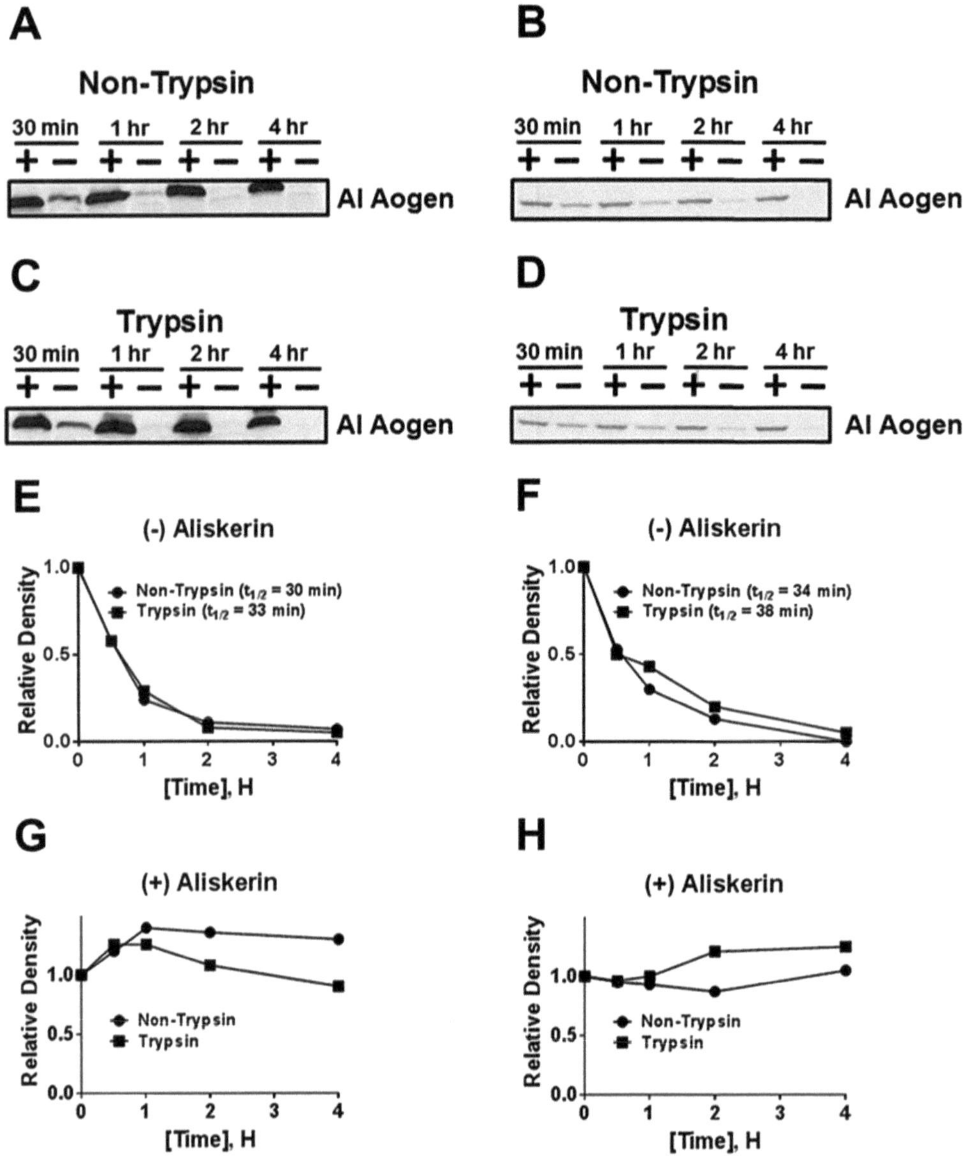

Fig. 3 Renin activity in the kidney mitochondria. Renin activity in the mitochondria of male (**a**, **c**, **e**, **g**) and female (**b**, **d**, **f**, **h**) sheep. Non-trypsin (**a**, **b**) or trypsin (**c**, **d**) treated mitochondria with (+) or without (−) the renin inhibitor aliskiren. Immunoblots were probed with an antibody directed to the Ang I sequence of angiotensinogen (AI-Aogen). The disappearance or half-life (t½) of AI-Aogen without (**e**, **f**) or with (**g**, **h**) aliskerin was determined by Prism 6 statistical program. Data adapted from Wilson et al., *Am J Physiol Renal Physiol* 310: F637–F645, 2016

The expression of renin in isolated mitochondria was confirmed by immunoblot analysis using two different renin antibodies. As shown in Fig. 4a and b, both antibodies detected two predominant bands at 37 and 55 kDa. The 37 kDa immunoreactive band corresponds to renin and the higher 55 kDa may represent "big renin,"

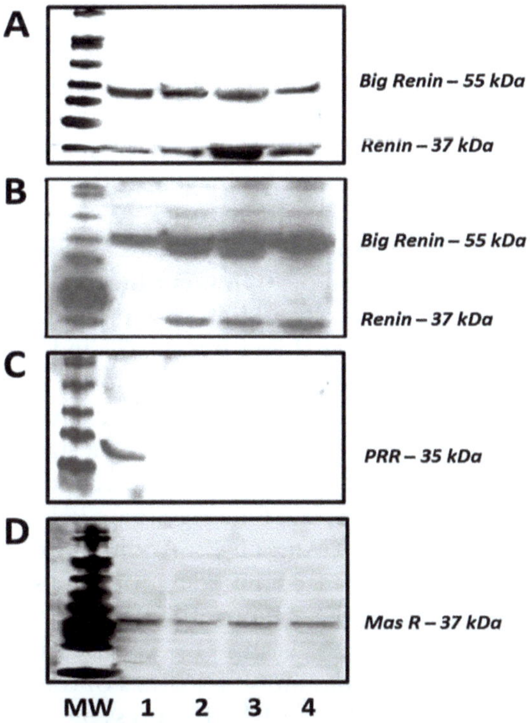

Fig. 4 Full-length immunoblot gels for renin, pro-renin receptor, and the AT_7R/Mas in isolated mitochondria. Renin protein detected by the Aviva (**a**) and Inagami (**b**) antibodies reveals two prominent bands at 37 and 55 kDa. A 35-kDa band is evident for the pro-renin receptor (PRR) in the renal cortex, but not in the mitochondrial fractions (**c**). The AT_7R/Mas protein visualized with the Alomone antibody reveals a single band at 37 kDa (**d**). Total cortical homogenate (*lane 1*) and three separate mitochondrial preparations (*lanes 2–4*). MW: molecular weight markers. Data adapted from Wilson et al., *Am J Physiol Renal Physiol* 310: F637–F645, 2016

an inactive form of the enzyme that is complexed to the renin-binding protein acetyl-glucosamine-2-epimerase [26]. We did not detect a protein band for the pro-renin receptor (PRR) in the isolated mitochondria, although a 35 kDa band was evident in the total cortical homogenate (Fig. 4c). The absence of the PRR in the mitochondria would appear to be consistent with lack of prorenin in this organelle. Finally, the immunoblot analysis revealed the Mas protein at ~37 kDa in three purified mitochondria preparations, as well as the total cortical homogenate (Fig. 4d). Using the same Alomone antibody, we previously demonstrated immunoreactive localization of the Mas protein to the proximal tubules, thick ascending limb, collecting ducts, and vasa recta of the sheep kidney [10].

Angiotensin peptides were detected within isolated mitochondria from the sheep renal cortex by distinct radioimmunoassays (RIAs) directed to the C-terminus of Ang I, Ang II and Ang-(1–7). Mitochondrial content of Ang II and Ang-(1–7) were both significantly higher than Ang I [23 ± 8 and 58 ± 17 vs. 2 ± 1 fmol/mg protein; respectively] [7]. Identification of angiotensin peptides within the mitochondria may portend for Ang-(1–7) and Ang II to influence mitochondrial function and integrity. Indeed Kim et al. found that Ang II induced oxidative stress, depolarization of the mitochondrial membrane potential, and cytosolic secretion of cytochrome C and apoptosis-inducing factor (AIF) in cultured rat proximal tubule cells [27]. Moreover, treatment with exogenous Ang-(1–7) attenuated the Ang II-induced mitochondrial production of ROS via reduced NOX4 expression and apoptosis [27].

Renal mitochondrial content of Ang I was markedly lower than Ang II and Ang-(1–7) and may reflect active processing within the mitochondria, therefore, we examined the potential processing pathways of Ang I in isolated mitochondrial homogenates using the ^{125}I-labeled peptide as a substrate coupled to HPLC-based separation and in-line γ-detection [2, 7, 28]. As shown in the chromatographs for Fig. 5a, ^{125}I-Ang I was converted primarily to ^{125}I-Ang-(1–7). Addition of the NEP inhibitor SCH39370 (SCH) reduced but did not abolish the ^{125}I-Ang-(1–7) peak (Fig. 5b). Co-addition of SCH and a TOP inhibitor (CPP) further reduced the ^{125}I-Ang-(1–7) peak and preserved the ^{125}I-Ang I peak (Fig. 5c). Immunoblot analysis of the isolated mitochondria confirmed both NEP and TOP expression (Fig. 5d). Quantification of Ang I metabolism revealed that the NEP inhibitor significantly increased Ang I and reduced Ang-(1–7) (Fig. 5e and f, respectively). The addition of the TOP inhibitor further reduced Ang-(1–7), but this did not attain significance compared with the effect of the NEP inhibitor alone. Importantly, we could not demonstrate the processing of Ang I to Ang II in the mitochondrial preparation either with or without the endopeptidase inhibitors (Fig. 5a–c). The failure to detect Ang II generation from Ang I may reflect the absence of ACE or other Ang II-generating enzymes in the isolated renal mitochondria. These data are consistent with the recent study by Astin et al. that did not detect ACE immunoreactivity within isolated liver mitochondria of the rat [29].

2 Materials

2.1 Tissue Collection

1. Renal Cortex: Sheep (*Ovis aries*): male and female, adult mixed breed (10–212 months of age).
2. Anesthesia: ketamine and isoflurane.
3. Ice-cold saline solution.

2.2 Preparation of Isolated Mitochondria

1. Freshly dissected sheep renal cortex tissue: 5–10 g.
2. Percoll.
3. Buffer A: 10 mM HEPES (Na^+ Free), 0.1 mM EDTA (Na^+), 1 L of MilliQ H_2O, pH = 7.4 (adjust with KOH only).

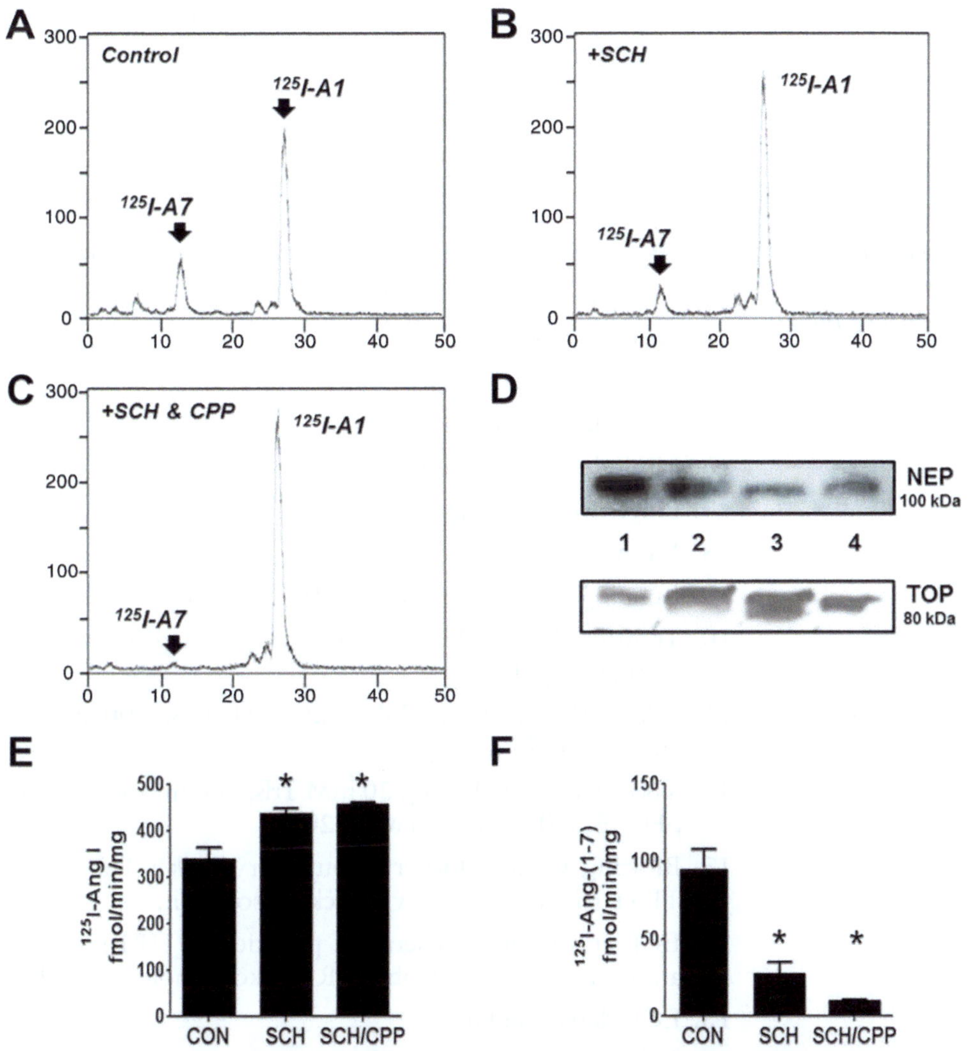

Fig. 5 Mitochondrial processing of ^{125}I-Ang I. Chromatograph reveals that ^{125}I-Ang I [^{125}I-AI] was converted to ^{125}I-Ang-(1–7) [^{125}I-A7] in renal mitochondrial homogenate fractions (**a**). ^{125}I-AI metabolism to ^{125}I-A7 was reduced by 10 μM SCH (**b**), and ^{125}I-AI metabolism to ^{125}I-A7 was reduced by 10 μM SCH in combination with 10 μM CPP (**c**). Immunoblot analysis revealed evidence for protein expression of neprilysin (NEP) and thimet oligopeptidase (TOP) in cortical homogenate (*lane 1*) and mitochondrial homogenates (*lanes 2–4*, $n = 3$) (**d**). ^{125}I-labeled products were separated by HPLC under gradient conditions. Statistical analysis of ^{125}I-AI metabolism demonstrated Scheme (10 μM) and SCH in combination with CPP (10 μM) preserved ^{125}I-Ang I, while Scheme (10 μM) and CPP (10 μM) reduced ^{125}I-Ang-(1–7) formation (**e**, **f**). The addition of SCH and CPP tended to further reduce ^{125}I-Ang-(1–7) and increase ^{125}I-Ang I (**e**, **f**). Data are mean ± SEM, $n = 3$; $*p < 0.05$ vs. control (CON). Data adapted from Wilson et al., *Am J Physiol Renal Physiol* 310: F637–F645, 2016

4. Homogenization buffer: 225 mM mannitol, 75 mM sucrose; dissolve the mannitol and sucrose in 200 mL of buffer A.
5. Percoll Buffer: 250 mM sucrose, 0.4 mM EDTA; dissolve the sucrose and EDTA in 200 mL of buffer A.

6. Polytron Ultra-Turrax T25 basic tissue homogenizer
7. 50 mL all-glass Dounce Homogenizer.
8. Nalgene round centrifuge tube, PC 16 mL.
9. 70-μm metal strainer.
10. Stir-Pak laboratory stirrer.

2.3 Immunoblot Analysis of Mitochondria Purity

1. Isolated sheep renal cortex mitochondria: 100–500 μg.
2. 2× Laemmli Sample Buffer.
3. Heating block.
4. 10% SDS-PAGE gels.
5. Mini-electrophoresis cell.
6. Mini-electrophoresis transfer cell.
7. Polyvinylidene difluoride (PVDF) membrane.
8. Methanol.
9. Running Buffer: 25 mM Tris, 192 mM glycine, 0.1% (w/v) SDS.
10. Transfer Buffer: 25 mM Tris, 192 mM glycine, 20% (v/v) methanol, pH 8.3.
11. Tris buffered saline (TBS): 20 mM Tris, 500 mM sodium chloride, pH 7.5.
12. TBS-Tween 20 (TBST): 20 mM Tris, 500 mM sodium chloride, pH 7.5, 0.05% (v/v) Tween 20.
13. Blotting Grade Blocker (Nonfat dry milk): This is added to TBST to make 5% (w/v) blocking solution.
14. ECL anti-mouse horseradish peroxidase (HRP) linked whole antibody, ECL anti-rabbit HRP linked whole antibody.
15. ECL detection kit.
16. X-ray/autoradiography film.
17. Antibodies:
    - Angiotensinogen, 1:1000, (Ang I directed, residues #25–34; Dr. Mark Chappell, WFUSM).
    - Angiotensinogen 1:1000, (Distal to Ang I sequence, residues #42–57; Dr. Mark Chappell, WFUSM).
    - ATP5A, UQCRC2, SDHB, 1:1000.
    - β-actin, 1:500 (#AC-15).
    - Cathepsin B, 1:400 (#CA10).
    - Mas receptor, 1:250 (#AAR-013).
    - Neprilysin, 1:2000 (#NEP11-S).
    - Prorenin receptor (ATP6IP2), 1:500 (#ab63957).

- Renin.
  - 1:3000 (#826; Dr. Tad Inagami, Vanderbilt University, Nashville TN).
  - 1:100 (#T100).
- Thimet oligopeptidase, 1:800 (#EPR7651).
- VDAC, 1:1000 (#4866).

2.4 Renin Assay

1. Isolated sheep renal cortex mitochondria: ~400 μg.
2. Nephrectomized sheep plasma.
3. Metabolism Buffer: 25 mM HEPES, 125 mM NaCl, 10 μM $ZnCl_2$ at pH 7.4 in MilliQ H_2O.
4. Trypsin, 5 mg/mL.
5. Soybean trypsin inhibitor (SBTI): 1 mg/mL.
6. EDTA.
7. *o*-Phenanthroline.

2.5 Angiotensin Peptide Quantification

1. Isolated sheep renal cortex mitochondria.
2. Milli-Q Ultrapore H_2O [≥18 MΩ].
3. Heptafluorobutyric acid (Sequanal Grade HFBA).
4. Tissue sonicator.
5. High-speed centrifuge up to 15,000 × *g*.
6. Methanol.
7. Sep-Pak C_{18} extraction columns.
8. Extraction manifold.
9. Radioimmunoassay (RIA) kits.
   - Angiotensin I.
   - Angiotensin II.
   - Angiotensin-(1–7)—in-house [29].

2.6 Angiotensin Peptide Metabolism

1. Isolated sheep renal cortex mitochondria: Mitochondrial lysate.
2. Metabolism buffer: 25 mM HEPES, 125 mM NaCl, 10 μM $ZnCl_2$ at pH 7.4 in MilliQ H_2O.
3. Iodine-125 (^{125}I) radionuclide (5 mCuries).
4. ^{125}I-Angiotensin I,
5. *N*-[*N*-[1-(*S*)-carboxyl-3-phenylpropyl]-(*S*)-phenyl-alanyl]-(*S*)-isoserine (SCH;10 μM): neprilysin inhibitor.
6. *N*-[1-(R,S)-carboxy-3-phenylpropyl]-AlaAla-Phe-*p*-aminobenzoate (CPP; 10 μM); thimet oligopeptidase inhibitor.
7. Shimadzu Prominence HPLC binary system.

2.7 HPLC Equipment

1. Aeris Peptide XB-C18 narrow bore column (2.1 × 100 mm).
2. Bioscan Inline γ-detector.
3. Phosphoric acid (HPLC grade).
4. Filter vials—0.45 μm PTFE, 450 μL capacity

3 Methods

3.1 Mitochondria Isolation from Renal Cortex

Procedure requires all steps performed on ice or at 4 °C

1. Dissect and weigh out 10 g of kidney cortex tissue, and mince very finely using surgical scissors and tissue grossing and trimming blade tools.
2. Split the 10 g of kidney tissue into two separate 15 mL conical tubes (5 g each tube).
3. Add homogenization buffer to each tube up to a total volume of 15 mL.

Homogenize on ice

1. Initially homogenize minced kidney utilizing a power blade metal tissue homogenizer. Then utilize the Teflon Dounce homogenizer (mortar and pestle) with 5 up and down strokes at setting of 5.
2. Filter homogenate through a 70 μm metal strainer.
3. Centrifuge tubes at 1500 × *g* for 5 min at 4 °C.
4. Remove supernatant and recentrifuge at 10,000 × *g* for 10 min at 4 °C to pellet the crude mitochondria.
5. Remove the supernatant and the pellet consists of the crude mitochondrial fraction.

Mitochondria crude pellet is resuspended in 12% Percoll media

1. Gently resuspend crude mitochondrial pellet in 12% Percoll media by diluting the 24% Percoll 1:1 (i.e., 3 mL of Percoll Buffer + 3 mL of 24% Percoll).

Layering the Percoll gradient

While a variety of tubes may be used for layering the Percoll gradient, we recommend using Nalgene Round Centrifuge Tube, PC 16 mL.

Percoll Gradient Calculations

24% Percoll media = 4.8 mL (100% Percoll) + 15.2 mL (Percoll Buffer) = 20 mL.

For a lesser volume of 24% Percoll media if preferred.

2.4 mL (100% Percoll) + 7.6 mL (Percoll Buffer) = 10 mL.

40% Percoll media = 6.0 mL (100% Percoll) + 9.0 mL (Percoll Buffer) = 15 mL.

1. Add 4 mL of 40% media to each tube. Then gently layer 5 mL of 24% Percoll media on top. For ease, even layering may be achieved by pipetting the 24% Percoll media in such a way that it sleeks slowly down the side of the tube. Finally, layer the 3 mL mitochondrial crude pellet in 12% Percoll media on top of 24% Percoll media layer.
2. Centrifuge at 15,000 × *g* for 5 min at 4 °C
3. Pure mitochondria will sediment with a prominent band between the 24% and 40% Percoll layers. Gently pipet off the mitochondrial band between the 24% and 40% interface from each tube and place into a clean Nalgene tube.
4. To dilute out the Percoll solution, add 8–10 mL of buffer A to each tube. Note: To further dilute the Percoll, divide the two tubes in half a second time. This will result in four tubes, and add ~4–5 mL of buffer A to each tube to further dilute the Percoll. Failure to efficiently dilute the Percoll media will prevent the sufficient formation of the mitochondrial pellet.
5. Centrifuge at 11,000 × *g* for 10 min at 4 °C.
6. Gently remove the supernatant so as not to disrupt the pure mitochondrial pellet.
7. Resuspend each mitochondrial pellet in preferred volume of metabolism buffer and transfer into 1.5 mL microfuge tubes and store at 4 °C or −80 °C for later use (Fig. 6).

3.2 Western Blot Analysis

1. Utilize purified mitochondrial preparations as described in Subheading 3.1.
2. Protein concentration of samples is determined by the Bradford assay (Bio-Rad).
3. A 2× Bio-Rad Laemmli buffer is added and mitochondrial samples are boiled in screw cap 1.5 mL tubes for 5 min. Equal volumes of samples are loaded onto a Bio-Rad MiniPROTEAN TGX StainFree 10% Gel.
4. Gels are run at 120 V until the dye front reaches the bottom of the gel.
5. Gels are removed from their plastic casts and are assembled on the semidry transfer cell. Two pieces of transfer buffer-soaked: filter paper and filter pad, methanol-activated PVDF transfer membrane are placed between the gel.
6. Ensure there are no air pockets trapped between the sandwich layers.
7. Transfer takes place for 60 min at 100 V in transfer buffer. Transfer cassette contains an ice pack to maintain temperature at 4 °C.

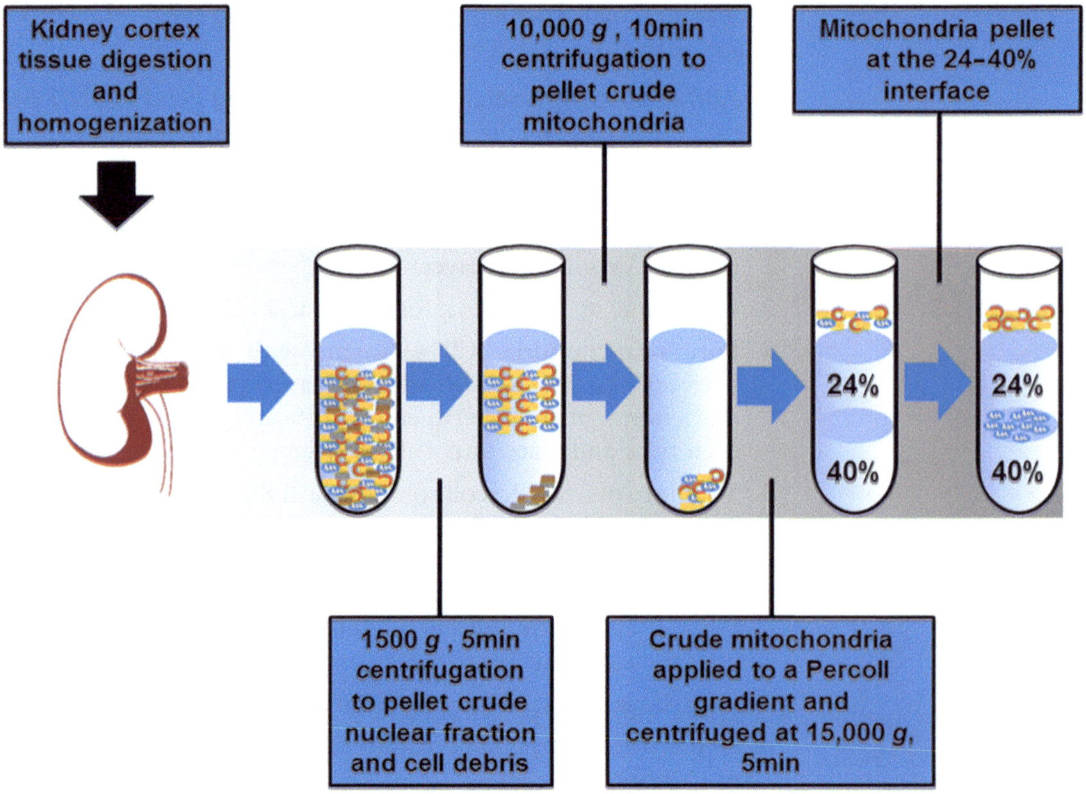

Fig. 6 Mitochondrial purification from renal cortex tissue. Mitochondria were isolated from fresh sheep renal cortex by a discontinuous Percoll gradient. Tissue was homogenized in mitochondrial homogenization buffer and centrifuged for 5 min at 1000 × *g*, and the resultant supernatant filtered through a 70 μm metal strainer and centrifuged for 10 min at 10,000 × *g*. This pellet, representing the crude mitochondrial fraction was resuspended in 12% Percoll media and gently layered onto a discontinuous Percoll gradient (24% and 40% in Percoll buffer). The gradient was centrifuged for 5 min at 15,000 × *g* and the band at the 24% and 40% Percoll layers was collected

8. The membrane is removed from the sandwich and is incubated with agitation at room temperature for 1 h in TBST supplemented with 5% Bio-Rad nonfat dry milk (w/v).
9. The buffer is removed and replaced by the same solution supplemented with the primary antibody. The membrane and antibody in buffer are agitated overnight at 4 °C.
10. The membrane is washed once for 10 min at room temperature in TBS and twice for 10 min at room temperature in TBST.
11. The membrane is incubated with agitation in TBST supplemented with 5% (w/v) Bio-Rad nonfat dry milk powder and the appropriate secondary antibody for 1–2 h at room temperature.
12. Wash the membrane as in **step 10**.

13. The membrane is finally incubated with ECL detection reagent, covered in a transparent plastic sheet protector, and exposed to X-ray film for an appropriate time. Films are subsequently developed in an automated developer.

3.3 Renin Assay

Mitochondria Reaction with Nephrectomized Sheep Plasma

1. We recommend utilizing ~400 μg of mitochondrial protein for the total reaction.
2. Trypsin activation (*see* **Note 1**).
3. Set up trypsin activation reaction mixture in a 1.5 mL tube by adding the reaction components listed sequentially in the table below and incubate for 1 h at 37 °C in a water bath (Table 1).
4. Stop the trypsin activation by adding the trypsin inhibitor SBTI (25 μL of 10 mg/mL stock solution and incubate at 37 °C for 15 min.
5. Divide the reaction volume by aliquoting 200 μL of total reaction volumes in separate 1.5 mL tubes. Label tubes by specifying "+" or "−" Aliskiren (Table 2).
6. Add aliskiren or metabolism buffer, mix and incubate for 10 min at 37 °C (*see* **Note 2–4**).
7. Add 10 μL of nephrectomized sheep plasma to each tube and vortex.

Table 1
Components of trypsin activation reaction mix for mitochondrial protein

Total reaction volume	400 μL
Metabolism buffer	335 μL
o-Phenanthroline (1 mM)	5 μL
EDTA (1 mM)	5 μL
Sheep mitochondria	50 μL
Trypsin (5 mg/mL)	5 μL

Table 2
Components of reaction mixture with and without the renin inhibitor, Aliskiren

(+) Aliskiren	(−) Aliskiren
200 μL—Total reaction mixture	200 μL—total reaction mixture
2 μL of (100 μM Aliskiren stock) = Final concentration 1 μM	2 μL of metabolism buffer

8. Remove 30 μL aliquots of (+) and (−) aliskiren at 30 min, 1 h, 2 h, and 4 h time points. When removing the aliquots at the specified time points, add the contents to a 1.5 mL screw cap tube and add 30 μL of 2× Bio-Rad Laemmli buffer.
9. Only 20 μL of the 60 μL total (30 μL Laemmli + 30 μL sample) will be loaded onto the gel for Western blot analysis (Fig. 7).
   Western Blot Gel Map (Table 3).
10. 20 μL of sample is loaded in each cell well [2–10]. Control sample (lane 2) was prepared by taking 10 μL of nephrectomized sheep plasma and diluting it in 200 μL of metabolism buffer. A 30 μL aliquot from the 210 μL (total volume) was removed and added to 30 μL of Laemmli buffer.
11. Immunoblot analysis is conducted with methods described in Subheading 3.2 utilizing the Ang I sequence of angiotensinogen antibody (Ang I-angiotensinogen, residues 25–34, 1:1000).

3.4 Angiotensin Peptide Quantification

1. After isolating mitochondria as described in Subheading 3.1, pellet mitochondria at 16,000 × *g* for 1 min.
2. Rapidly dilute the sonicated mitochondria in 10 mL of MilliQ H_2O in a 15 mL plastic conical tube. Puncture the cap of the tube.

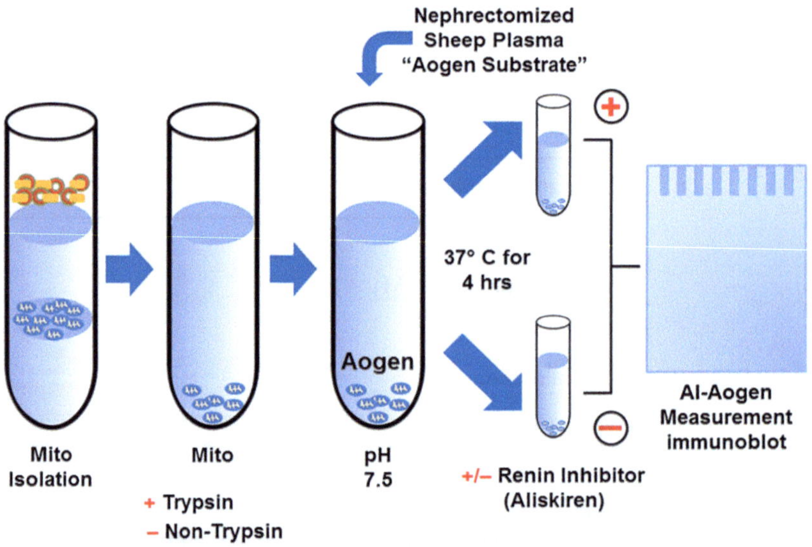

Fig. 7 Renin assay utilizing isolated sheep renal cortex mitochondrial homogenates. Mitochondria were isolated as described in Fig. 6. The mitochondrial pellet was reconstituted in 1 mL of metabolism buffer and stored on ice. For basal renin activity assay, 400 μg of mitochondrial homogenate was added to 10 μL of nephrectomized sheep plasma as the source of exogenous angiotensinogen (Aogen) substrate. The assay was performed at 37 °C for 4 h in the presence or absence of the renin inhibitor aliskiren (1 μM, final concentration). Aliquots were removed at 30 min, 1, 2 and 4 h and analyzed by Western blot using the Ang I-angiotensinogen (AI-Aogen) antibody. Activation of prorenin was performed by addition of 5 mg/mL trypsin to the mitochondrial sample for 60 min. Trypsin was inactivated by incubation with an excess of soybean trypsin inhibitor (1 mg/mL) for 15 min at room temperature

Table 3
Gel map for Western blotting mitochondrial angiotensinogen and other proteins of the RAS

1	2	3	4	5	6	7	8	9	10
Molecular weight marker	Control Neph Sheep Plasma	30 min + Aliskiren	30 min − Aliskiren	1 h + Aliskiren	1 h − Aliskiren	2 h + Aliskiren	2 h − Aliskiren	4 h + Aliskiren	4 h − Aliskiren

3. Immerse tube in boiling water for 15 min.
4. Sonicate the mitochondrial homogenate ~30 s at a setting of 5.
5. Remove an aliquot for protein analysis by Bradford (100 μL).
6. Add 10 μL of HFBA to the mixture to acidify the solution (0.1% HFBA final concentration).
7. Centrifuge at 10,000 × *g* for 20 min at 4 °C.
8. Collect supernatant and transfer to a clean 15 mL conical tube. Discard pellet.
9. Place supernatant sample on ice until SepPak extraction.

Angiotensin Peptide Extraction

10. Peptide extraction is performed on Sep-Pak C_{18} columns coupled to 10 mL BD plastic syringes. The syringe fits snugly on the column and is ideal as a 10 mL reservoir. The acidified supernatant is applied to the syringe and is gravity fed through the Sep-Pak column (*see* Fig. 8).
11. Activate Sep Pak (300 mg cartridges; Millipore, MA):
    - 5 mL 100% methanol–0.1% HFBA.
    - 5 mL 0.1% HFBA in MilliQ H_2O.
12. Apply samples (in 0.1% HFBA/MilliQ H_2O), gravity feed through the syringe reservoir.
13. Wash:
    - 10 mL 0.1% HFBA in MilliQ H_2O.
    - 10 mL MilliQ H_2O.
14. Elute in 3 mL 100% methanol–0.1% HFBA gravity feed.
15. Aliquot specified volume for RIA analysis. It is important to completely evaporate samples in a vacuum centrifuge as any residual solvent may drastically interfere with the RIA analysis (*see* **Notes 8–11**).

3.5 Angiotensin Peptide Metabolism

Reaction Vessel—Example in 1.5 mL tube (Table 4)

1. In a 1.5 mL reaction tube, combine reaction contents in the following order.

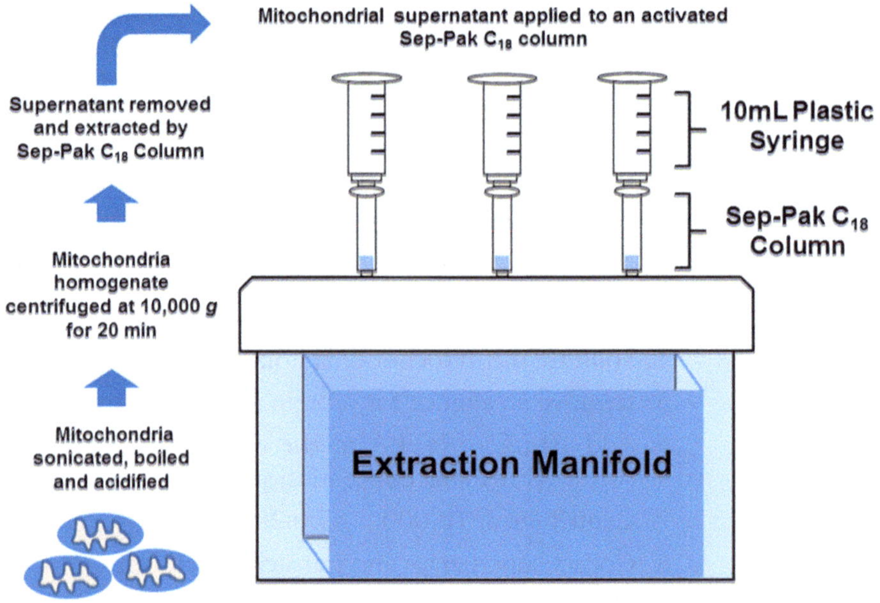

Fig. 8 Angiotensin peptide extraction from isolated renal cortex mitochondria. Sheep renal cortex mitochondria were isolated as described in Fig. 6. The mitochondrial pellet was reconstituted in Milli-Q water and placed in a boiling water bath for 15 min. The mitochondrial fraction was acidified with heptafluorobutyric acid (HFBA) to a final concentration of 0.1%, sonicated and centrifuged at 10,000 × *g* for 20 min at 4 °C. The resultant supernatant was applied to an activated Sep-Pak C_{18} extraction column, washed with 0.1% HFBA, and the peptide fraction eluted with 3 mL of 100% methanol in 0.1% HFBA. Measurement of immunoreactive Ang I, Ang II, and Ang-(1–7) in the extracted mitochondria was performed using three separate radioimmunoassays (RIAs)

Table 4
Reaction mix for measuring mitochondrial angiotensin peptides

Total reaction volume	500 μL
Metabolism buffer—Volume will vary based on protein concentration of sample	390 μL
Inhibitor of choice prepared at a 10× higher concentration	50 μL
Mito sample (~50 μg)—Volume will vary based on protein concentration of sample	10 μL
^{125}I-Ang peptide (2 × 10^6 cpm/100 μL) in 0.1% Triton—Volume may vary based on the cpm of radioactivity	50 μL

2. Add precalculated volume of metabolism buffer.
3. Add inhibitors at desired concentration.
4. Add mitochondria sample in desired protein concentration and pre-incubate for ~5 min.
5. Add ^{125}I-labeled peptide substrate and vortex.
6. Incubate reaction at 37 °C for the predetermined time period (i.e., 15, 30, or 60 min).

7. At every time point, vortex each tube and immediately pipette 220 μL of suspension into a tube containing 220 μL of 1% phosphoric acid to terminate the metabolism reaction.
8. Repeat for each reaction sample sequentially.
9. Vortex aliquots and centrifuge at 16,500 × *g* for 1 min.
10. Filter samples using HPLC grade Filter Vials (0.45 μm PTFE, 450 μL capacity).
11. Refrigerate samples until submission for HPLC analysis (Fig. 9).
12. HPLC conditions for separation of ^{125}I-peptides are described below (*see* **Notes 11–14**):
    - Solvent A: 0.1% phosphoric acid in Milli-Q ultrapure water.
    - Solvent B: 80% acetonitrile–0.1% phosphoric acid.
    - Flow rate: 0.35 mil/min.
    - Gradient: Isocratic 5 min at 10% B; Liner 10–30% B; Isocratic 30% B 10 min.

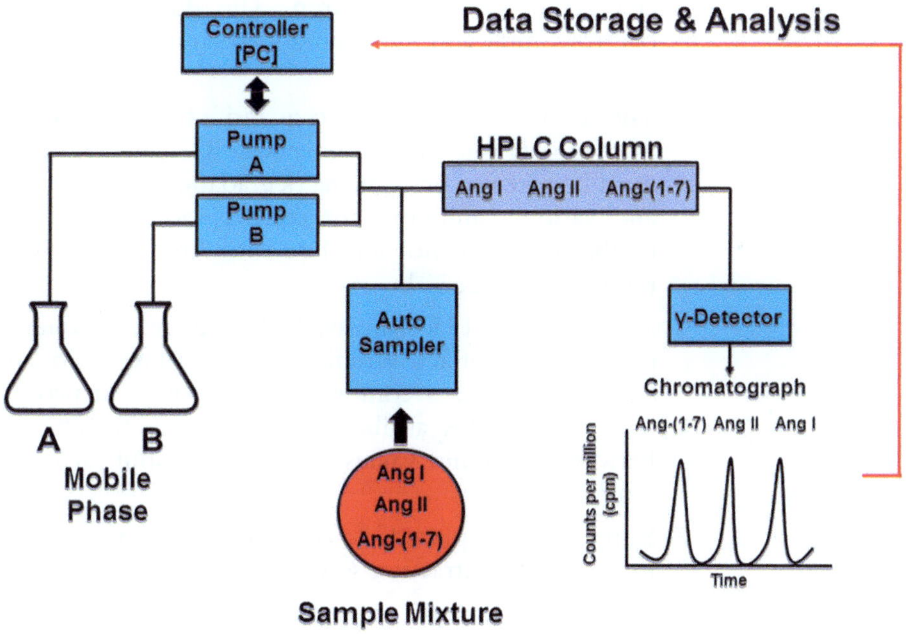

Fig. 9 Scheme for HPLC-based detection of ^{125}I-Ang I metabolism in isolated renal cortex mitochondria. ^{125}I-Ang I metabolism is measured at 37 °C in the mitochondrial lysate. The metabolism reaction contains a final concentration of 0.5 nM ^{125}I-Ang I and 100 nM unlabeled Ang I with or without peptidase inhibitors. The reaction is stopped with ice-cold 1.0% phosphoric acid and centrifuged at 16,000 × *g*. The supernatants are filtered by filter vials, and the products are separated by C18 reverse phase chromatography. The individual radioactive peaks separated by the HPLC column are detected by an inline γ detector

4 Notes

1. The rationale for trypsin activation is that renin is initially synthesized in a pro-form that is inactive towards the angiotensinogen substrate. Upon enzymatic cleavage of the pro-segment by serine proteases such as trypsin, renin is fully active and hydrolyzes the Ang I segment from angiotensinogen.
2. Nephrectomized sheep plasma is used as an exogenous source of angiotensinogen. The removal of the kidneys abolishes the major source of circulating pro-renin and renin. As a result of the essentially complete depletion of renin, circulating levels of intact angiotensinogen rise markedly due to the inhibition of negative feedback by the lower levels of Ang II. Therefore, the plasma isolated from nephrectomized sheep blood constitutes an enriched source of intact angiotensinogen for experimental studies.
3. Renin is a highly specific protease and the sole endogenous substrate for the enzyme is angiotensinogen. Sheep renin will readiy hydrolyze angiotensinogen from other species [30].
4. To distinguish renin activity in the mitochondria from that of other aspartyl proteases, we pretreat the sample with the specific renin inhibitor aliskiren. The difference between the control sample and the renin-added sample is considered renin activity.
5. The classical renin assay is to quantify the generation of Ang I from angiotensinogen using a specific RIA or ELISA. Precautions must be taken to prevent the degradation of Ang I in the assay and various inhibitors are typically added; however, these inhibitors should not interfere with Ang I assay or attenuate renin activity [2, 30].
6. Alternatively, ELISAs are available to quantify immunoreactive protein levels of active renin. To detect pro-renin, aliskerin is added to induce a conformational change in pro-renin and allow the ELISA antibody to detect renin. The difference between the renin values with and without aliskerin is pro-renin [2, 30].
7. Samples can be placed in a desiccator under vacuum overnight to ensure the extraction solvents are completely removed.
8. RIA analysis requires an appropriate "blank sample" that lacks the mitochondria but is extracted in the exact same manner and assayed under identical conditions.
9. The recovery of angiotensins during the extraction process can be determined by adding a small amount of the radiolabeled

peptide to each sample homogenate. The % recovery is calculated by dividing the recovered cpm by the initial cpm × 100; typical recovery values vary from 60% to 90%. Peptide content can be corrected for each sample by normalizing to the recovery value. Alternatively, an excess of unlabeled peptide is added to a duplicate set of homogenate samples and the recovery is based on this value.

10. The RIA or ELISA peptides values can be further validated by prior separation of the sample on HPLC and the collected fractions assayed by the RIA or ELISA. Alternatively, samples can be fractionated by HPLC and the peptides quantified by mass-spectroscopy.
11. The metabolites are identified by comparison of their retention times to the ^{125}I-standard peptides, the response to the addition of peptidase inhibitors and immunoprecipitation by selective antibodies. ^{125}I-standards are developed using the chloramine T method to iodinate the tyrosine residue of angiotensin peptides and purify the radiolabeled peptides by HPLC to a specific activity of 2200 Curies/mmol [2, 28].
12. Radioactive peaks are quantified based on the specific activity of the ^{125}I-peptide substrate [~2200 Curies/mmol]. In brief, 1 Curie = 2.22×10^{12} dpm; the sample peak cpm is converted to dpm based on the efficiency of the γ-gamma counter (~70%) and divided by the specific activity to yield the mole content.
13. The Shimadzu auto-sampler employs an extensive wash cycle between sample injections to eliminate the radioactive carryover from the previous samples.
14. This method detects only the ^{125}I-tyrosine signature and metabolites that do not contain the ^{125}I-tyrosine remain unidentified [2, 28].

Acknowledgment

The authors gratefully acknowledge Nancy Pirro, Eric LeSaine, and Pamela Dean for technical and surgical support. During the completion of this work, B.A. Wilson was supported by an American Heart Association (AHA) predoctoral fellowship grant (15PRE25120007). Additional support for these studies was provided by National Institutes of Health Grants HD-047584, HD-017644, HD-084227, HL-51952, T32 grant (HL091797); the Groskert Heart Fund, the Wake Forest Venture Fund, and the Farley-Hudson Foundation (Jacksonville, NC).

References

1. Campbell DJ (2014) Clinical relevance of local renin angiotensin systems. Front Endocrinol (Lausanne) 5:113. doi:10.3389/fendo.2014.00113
2. Chappell MC (2016) Biochemical evaluation of the renin-angiotensin system: the good, bad, and absolute? Am J Physiol Heart Circ Physiol 310(2):H137–H152. doi:10.1152/ajpheart.00618.2015
3. Santos RA (2014) Angiotensin-(1–7). Hypertension63(6):1138–1147.doi:10.1161/HYPERTENSIONAHA.113.01274
4. Abadir PM, Foster DB, Crow M, Cooke CA, Rucker JJ, Jain A, Smith BJ, Burks TN, Cohn RD, Fedarko NS, Carey RM, O'Rourke B, Walston JD (2011) Identification and characterization of a functional mitochondrial angiotensin system. Proc Natl Acad Sci U S A 108(36):14849–14854. doi:10.1073/pnas.1101507108
5. Gwathmey TM, Shaltout HA, Pendergrass KD, Pirro NT, Figueroa JP, Rose JC, Diz DI, Chappell MC (2009) Nuclear angiotensin II type 2 (AT2) receptors are functionally linked to nitric oxide production. Am J Physiol Renal Physiol 296(6):F1484–F1493. doi:10.1152/ajprenal.90766.2008
6. Pendergrass KD, Gwathmey TM, Michalek RD, Grayson JM, Chappell MC (2009) The angiotensin II-AT1 receptor stimulates reactive oxygen species within the cell nucleus. Biochem Biophys Res Commun 384(2):149–154. doi:10.1016/j.bbrc.2009.04.126
7. Wilson BA, Nautiyal M, Gwathmey TM, Rose JC, Chappell MC (2015) Evidence for a mitochondrial angiotensin-(1–7) system in the kidney. Am J Physiol Renal Physiol 00479:02015. doi:10.1152/ajprenal.00479.2015
8. Abadir PM, Walston JD, Carey RM (2012) Subcellular characteristics of functional intracellular renin-angiotensin systems. Peptides 38(2):437–445. doi:10.1016/j.peptides.2012.09.016
9. De Mello WC (1998) Intracellular angiotensin II regulates the inward calcium current in cardiac myocytes. Hypertension 32(6):976–982
10. Gwathmey TM, Pendergrass KD, Reid SD, Rose JC, Diz DI, Chappell MC (2010) Angiotensin-(1–7)-angiotensin-converting enzyme 2 attenuates reactive oxygen species formation to angiotensin II within the cell nucleus. Hypertension 55(1):166–171. doi:10.1161/HYPERTENSIONAHA.109.141622
11. Gwathmey TM, Westwood BM, Pirro NT, Tang L, Rose JC, Diz DI, Chappell MC (2010) Nuclear angiotensin-(1–7) receptor is functionally coupled to the formation of nitric oxide. Am J Physiol Renal Physiol 299(5):F983–F990. doi:10.1152/ajprenal.00371.2010
12. Kumar R, Thomas CM, Yong QC, Chen W, Baker KM (2012) The intracrine renin-angiotensin system. Clin Sci (Lond) 123(5):273–284. doi:10.1042/CS20120089
13. Kurdi M, De Mello WC, Booz GW (2005) Working outside the system: an update on the unconventional behavior of the renin-angiotensin system components. Int J Biochem Cell Biol 37(7):1357–1367. doi:10.1016/j.biocel.2005.01.012
14. Re RN, Cook JL (2011) Noncanonical intracrine action. J Am Soc Hypertens 5(6):435–448. doi:10.1016/j.jash.2011.07.001
15. de Cavanagh EM, Inserra F, Ferder M, Ferder L (2007) From mitochondria to disease: role of the renin-angiotensin system. Am J Nephrol 27(6):545–553. doi:10.1159/000107757
16. Haller H, Lindschau C, Erdmann B, Quass P, Luft FC (1996) Effects of intracellular angiotensin II in vascular smooth muscle cells. Circ Res 79(4):765–772
17. Zhuo JL, Li XC, Garvin JL, Navar LG, Carretero OA (2006) Intracellular ANG II induces cytosolic Ca2+ mobilization by stimulating intracellular AT1 receptors in proximal tubule cells. Am J Physiol Renal Physiol 290(6):F1382–F1390. doi:10.1152/ajprenal.00269.2005
18. Li XC, Zhuo JL (2008) Intracellular ANG II directly induces in vitro transcription of TGF-beta1, MCP-1, and NHE-3 mRNAs in isolated rat renal cortical nuclei via activation of nuclear AT1a receptors. Am J Physiol Cell Physiol 294(4):C1034–C1045. doi:10.1152/ajpcell.00432.2007
19. de Cavanagh EM, Piotrkowski B, Basso N, Stella I, Inserra F, Ferder L, Fraga CG (2003) Enalapril and losartan attenuate mitochondrial dysfunction in aged rats. FASEB J 17(9):1096–1098. doi:10.1096/fj.02-0063fje
20. Nautiyal M, Katakam PV, Busija DW, Gallagher PE, Tallant EA, Chappell MC, Diz DI (2012) Differences in oxidative stress status and expression of MKP-1 in dorsal medulla of transgenic rats with altered brain renin-angiotensin system. Am J Physiol Regul Integr Comp Physiol 303(8):R799–R806. doi:10.1152/ajpregu.00566.2011
21. Rajapakse N, Shimizu K, Payne M, Busija D (2001) Isolation and characterization of intact mitochondria from neonatal rat brain. Brain Res Brain Res Protoc 8(3):176–183

22. Taugner R, Buhrle CP, Nobiling R, Kirschke H (1985) Coexistence of renin and cathepsin B in epithelioid cell secretory granules. Histochemistry 83(2):103–108
23. Clausmeyer S, Sturzebecher R, Peters J (1999) An alternative transcript of the rat renin gene can result in a truncated prorenin that is transported into adrenal mitochondria. Circ Res 84(3):337–344
24. Wanka H, Kessler N, Ellmer J, Endlich N, Peters BS, Clausmeyer S, Peters J (2009) Cytosolic renin is targeted to mitochondria and induces apoptosis in H9c2 rat cardiomyoblasts. J Cell Mol Med 13(9A):2926–2937. doi:10.1111/j.1582-4934.2008.00448.x
25. Ishigami T, Kino T, Chen L, Minegishi S, Araki N, Umemura M, Abe K, Sasaki R, Yamana H, Umemura S (2014) Identification of bona fide alternative renin transcripts expressed along cortical tubules and potential roles in promoting insulin resistance in vivo without significant plasma renin activity elevation. Hypertension 64(1):125–133. doi:10.1161/HYPERTENSIONAHA.114.03394
26. Takahashi S, Hori K, Ogasawara H, Hiwatashi K, Sugiyama T (2006) Effects of nucleotides on the interaction of renin with GlcNAc 2-epimerase (renin binding protein, RnBP). J Biochem 140(5):725–730. doi:10.1093/jb/mvj201
27. Kim SM, Kim YG, Jeong KH, Lee SH, Lee TW, Ihm CG, Moon JY (2012) Angiotensin II-induced mitochondrial Nox4 is a major endogenous source of oxidative stress in kidney tubular cells. PLoS One 7(7):e39739. doi:10.1371/journal.pone.0039739
28. Shaltout HA, Westwood BM, Averill DB, Ferrario CM, Figueroa JP, Diz DI, Rose JC, Chappell MC (2007) Angiotensin metabolism in renal proximal tubules, urine, and serum of sheep: evidence for ACE2-dependent processing of angiotensin II. Am J Physiol Renal Physiol 292(1):F82–F91. doi:10.1152/ajprenal.00139.2006
29. Astin R, Bentham R, Djafarzadeh S, Horscroft JA, Kuc RE, Leung PS, Skipworth JR, Vicencio JM, Davenport AP, Murray AJ, Takala J, Jakob SM, Montgomery H, Szabadkai G (2013) No evidence for a local renin-angiotensin system in liver mitochondria. Sci Rep 3:2467. doi:10.1038/srep02467
30. Campbell DJ, Nussberger J, Stowasser M, Danser AH, Morganti A, Frandsen E, Menard J (2009) Activity assays and immunoassays for plasma renin and prorenin: information provided and precautions necessary for accurate measurement. Clin Chem 55(5):867–877. doi:10.1373/clinchem.2008.118000

Chapter 10

Comprehensive Assessments of Energy Balance in Mice

Justin L. Grobe

Abstract

Increasing evidence supports a major role for the renin-angiotensin system (RAS) in energy balance physiology. The RAS exists as a circulating system but also as a local paracrine/autocrine signaling mechanism in target tissues including the gastrointestinal tract, the brain, the kidney, and distinct adipose beds. Through activation of various receptors in these target tissues, the RAS contributes to the control of food intake behavior, digestive efficiency, spontaneous physical activity, and aerobic and anaerobic resting metabolism. Although the assortment of methodologies available to assess the various aspects of energy balance can be daunting for an investigator new to this area, a relatively straightforward array of entry-level and advanced methodologies can be employed to comprehensively and quantitatively dissect the effects of experimental manipulations on energy homeostasis. Such methodologies and a simple initial workflow for the use of these methods are described in this chapter, including the use of metabolic caging systems, bomb calorimetry, body composition analyzers, respirometry systems, and direct calorimetry systems. Finally, a brief discussion of the statistical analyses of metabolic data is included.

Key words Energy balance, Caloric balance, Obesity, Calorimetry, Metabolism, Digestive efficiency, Energy efficiency, Resting metabolic rate, Aerobic, Anaerobic

1 Introduction

Genetic, dietary, behavioral, and pharmacological manipulations often result in changes in energy balance. Such changes are primarily mediated through altered food intake behavior, digestive efficiency, spontaneous physical activity, and aerobic or anaerobic resting metabolism (Table 1). As we recently reviewed, the renin-angiotensin system (RAS) has been implicated in the physiological/pathophysiological control of each of these endpoints [1]. Intriguingly, the RAS can simultaneously activate opposing effects (depending upon the tissue site of action, receptors activated, and endpoint of interest) which can be reflected in large and opposing changes in individual phenotypes (e.g., food intake and resting metabolic rate) that ultimately result in no net change in body mass. Thus, investigators studying the effects of various manipulations of the RAS upon metabolic physiology may benefit

Sean E. Thatcher (ed.), *The Renin-Angiotensin-Aldosterone System: Methods and Protocols*, Methods in Molecular Biology, vol. 1614, DOI 10.1007/978-1-4939-7030-8_10, © Springer Science+Business Media LLC 2017

Table 1
The four major mechanisms of energy flux, and common methodologies to assess these components

	Energy intake mechanisms		Energy output mechanisms	
	Food intake	Digestive efficiency	Physical activity	Resting metabolism
Type of mechanism	Behavioral	Biological	Behavioral	Biological
Contribution to energy balance	Maximally sets the amount of energy to be balanced by expenditure	Scales energy (calories) actually absorbed from 0 to 100% of those consumed	Typically contributes 0–40% of total energy expenditure	Typically contributes 60–100% of total energy expenditure
Common methods	Food weights (quantitative), consumption patterning (qualitative)	Bomb calorimetry (quantitative), fecal acid steatocrit or metabolomics (qualitative)	Photoelectric beam break, wheel rotations, radiotelemetric implant (all qualitative)	Direct calorimetry (for total metabolic rate), respirometry (aerobic metabolism) (all quantitative)

from a simple rubric for dissecting the relative effects of the RAS upon individual mechanisms that contribute to energy balance in addition to the complex, integrated endpoint of total body mass.

It is important to recognize that at the population level, human obesity is the result of very small net imbalances of caloric intake versus output spread over long periods of time. One highly publicized estimate suggests that roughly 7.2 kcal/day imbalance between input and output can explain the obesity epidemic in the USA [2]. Compared to a normal 2000 kcal/day diet, 7.2 kcal/day reflects a 0.35% imbalance. While the exact accuracy of this estimate has been contested by various groups, there is essentially universal agreement that obesity is the result of *very* small differences between caloric input and output. By extension, any physiological or behavioral mechanisms that contribute to energy balance must be very tightly controlled. Therefore, all mechanisms contributing to energy balance represent potential therapeutic targets for obesity and stopping one's investigation of the effects of any manipulation (of the RAS or any system) after only measuring body mass and food intake behaviors is inadequate and inappropriate. At minimum, investigators should consider calculating the fraction of an observed change in body mass that can be attributed to a specific endpoint such as altered food intake behavior.

In this chapter, we outline a simple workflow (Fig. 1) which uses a combination of classic and emerging technologies to allow for straightforward, comprehensive, and quantitative assessments

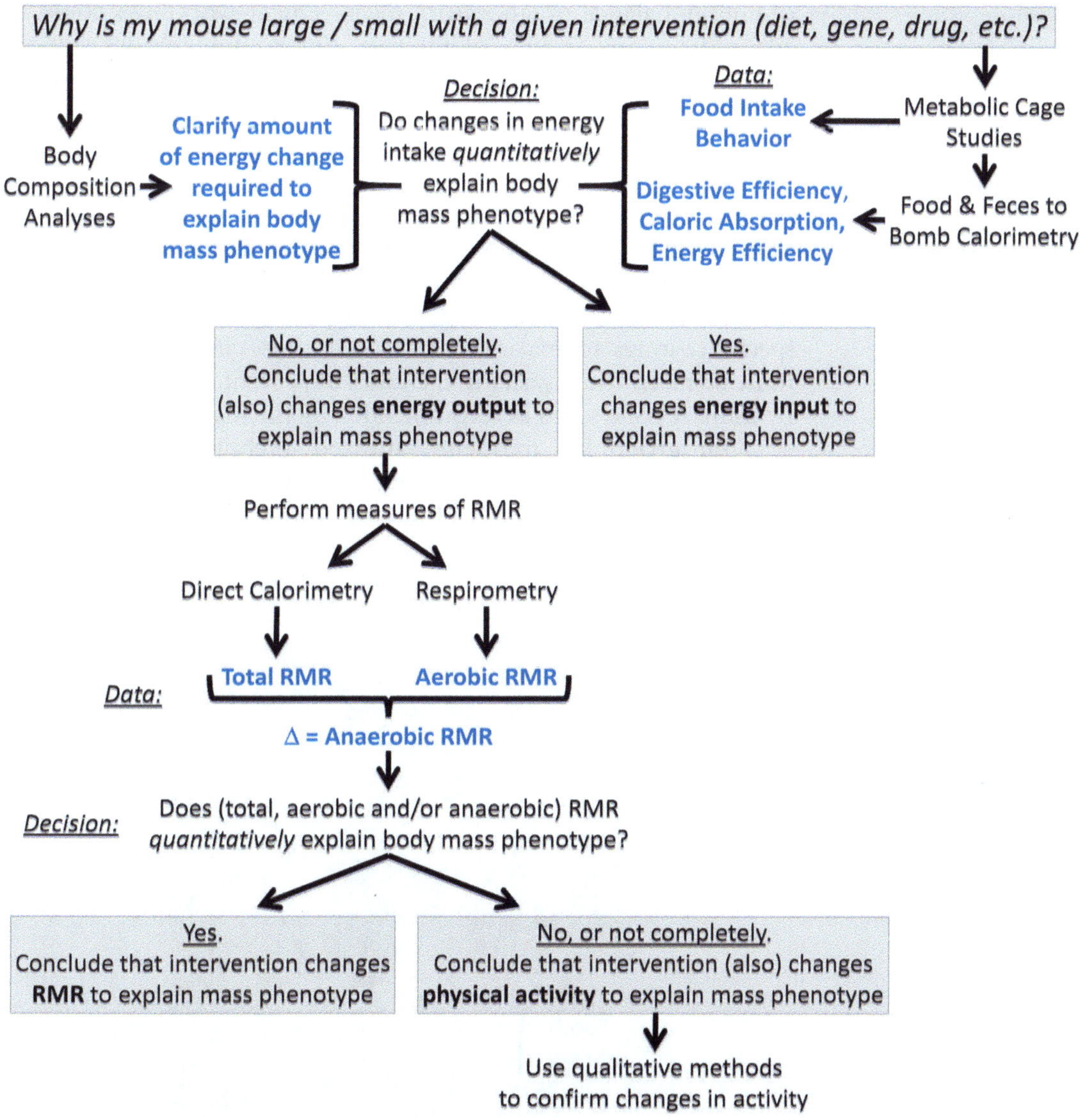

Fig. 1 Proposed workflow to assess energy balance in laboratory mice. To comprehensively identify the mechanism(s) that contribute to an observed body mass phenotype, a quantitative approach is required. One simple workflow is to use body composition analysis to clarify the total amount of energy change that is required to produce an observed metabolic phenotype, then sequentially—and *quantitatively*—determine the relative contributions of the four major mechanisms of energy balance to this total. Herein we propose that the use of metabolic caging systems and bomb calorimetry can quantify the contributions of food intake behavior and digestive efficiency (and in combination, caloric intake); if these mechanisms account for essentially all of the total change in energy flux, then the investigator can conclude that caloric intake changes explain the overall phenotype. If these mechanisms do not quantitatively account for the total energy change, then subsequent studies to explore resting metabolic rate (RMR) are required. Only after an investigator exhausts methods to assess RMR can changes in physical activity be quantitatively determined. Qualitative methods to assess physical activity are then recommended to complement the preceding quantitative methods

of energy balance in laboratory mice. Similar workflows could be adapted for any species of interest, but equipment descriptions herein are tailored for use with mice.

2 Materials

2.1 Metabolic Caging Systems

One common type of metabolic caging system for use with mice is the Nalgene-style single-mouse sized metabolic cage. These cages are designed to be used with a single mouse per cage, and are treated with special coatings to maximize urine collection efficiency. Great care must be taken when washing the cages to prevent loss of the coatings by using specific soaps as recommended by the manufacturer.

Our group has frequently utilized Techniplast USA model 3600M021 to study the effect of manipulations of the RAS upon metabolic and fluid balance physiology [3–8] (Fig. 2). These cages have been modified to replace the standard water bottle with two 50 mL burets, to allow for two-bottle choice testing.

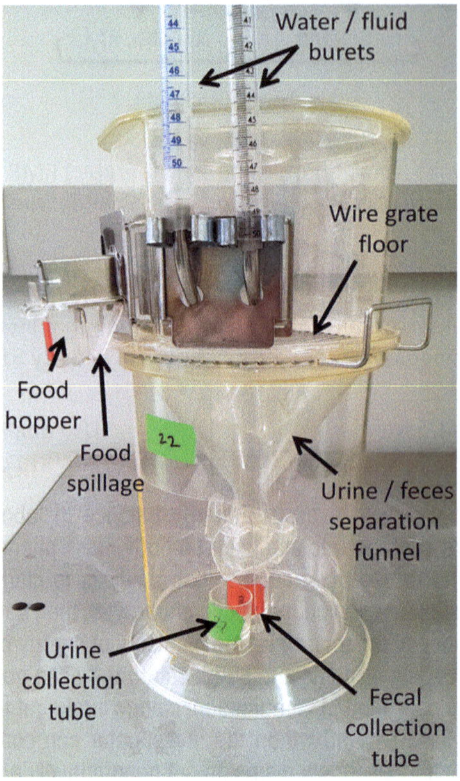

Fig. 2 Illustration of one type of mouse-sized metabolic cage. A single mouse-sized metabolic cage (Techniplast USA model 3600 M021 shown) typically includes a detachable food hopper, water bottle, urine and fecal collection tubes. Cages can be modified (as shown) to provide multiple drink tubes, to permit two-bottle choice tests

2.2 Bomb Calorimeter

A bomb calorimeter of appropriate size must be selected. One commonly used design is a 25–200 mg bomb (e.g., Parr Instrument Company, model 6725). Importantly the investigator will need a pellet press to uniformly prepare samples for use in the bomb (e.g., Parr Instrument Company, model 2817). We have used these models of equipment to examine digestive efficiency and energy efficiency in animals with manipulations of mitochondrial function and the gut microbiome [9, 10] (Fig. 3).

2.3 Body Composition Analyzer

Body composition analyzers are typically far beyond the budget of an individual laboratory, and are more typically offered through an institutional core facility. It is important to know what equipment is available in the facility, however, as occasionally the resolution of a given scanner (e.g., a rat-sized scanner) is insufficient to accurately assess body composition of smaller species (e.g., mice). We have used NMR and MRI scanners within core facilities at the University of Iowa in a wide array of studies of energy balance [3, 8, 10–12]. Protocols for using such equipment will be provided by core facility staff and are beyond the scope of this chapter. These analyzers typically provide bodily composition of fat mass, lean mass, and water.

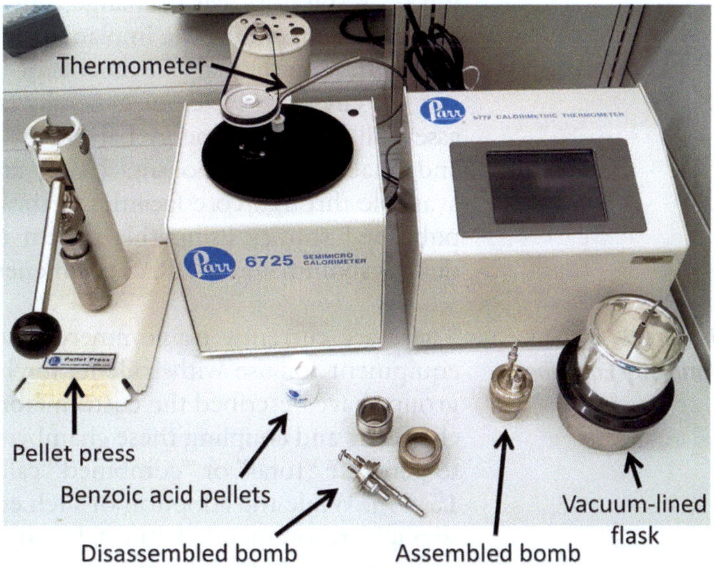

Fig. 3 Illustration of a semi-micro bomb calorimetry system. A 25–200 mg bomb calorimeter system (Parr model 6725 shown) consists of a bomb that is loaded with the sample pellets (prepared by a press as shown), which is then submerged in a specific volume of water within a vacuum-lined flask. The flask is then placed into the mixing chamber that includes a high-resolution thermometer in the water bath. Systems such as the Parr model 6725 perform scripted analysis runs (roughly 20 min per sample) and also log and analyze data for the user

2.4 Respirometry Equipment

Gas analyzer equipment is available from a long list of manufacturers, including AEI Technologies, Sable Systems, and Columbus Instruments. It is again important to evaluate the specifications of the equipment under comparison based on the resolution of the analyzers for the species of animal that will be studied. Equipment to correct air flow rates to standard temperature and pressure is available from many commercial sources as well. Some manufacturers offer equipment to record data, while others require that you interface these systems with third-party chart recorder equipment (e.g., ADInstruments). Finally, housing chambers can be obtained from some manufacturers, or custom-built. Investigators who wish to be able to expose animals to a range of ambient temperatures may consider the use of custom-modified water-jacketed chambers coupled to circulating heating/cooling water baths to provide such options.

We have described a push-pull respirometry array using AEI gas analyzers and custom thermally controlled chambers (ACE Glass, Vineyard, NJ) in multiple recent publications [3, 5, 7, 8, 13, 14] (Fig. 4).

2.5 Comprehensive Phenotyping Systems

Several companies offer "comprehensive" phenotyping equipment (e.g., "OxyMax/CLAMS" from Columbus Instruments, "Promethion" from Sable Systems, "PhenoMaster" from TSE). These combinatorial systems allow simultaneous assessment of food intake, respiratory gas exchange, physical activity by beam break, and core temperature via implanted telemeters. Such systems offer repeated, longitudinal assessments of various phenotypes throughout the circadian cycle but generally fail to include mechanisms to assess all aspects of energy balance by ignoring digestive efficiency and anaerobic metabolism. Using an OxyMax/CLAMS system available through core facilities at the University of Iowa, we have published studies using this system to investigate the effects of dietary sodium and the RAS upon energy balance [8].

2.6 Direct Calorimetry Equipment

There are currently no commercial sources of direct calorimetry equipment for use with rodents for biomedical research. Various groups have described the custom fabrication of direct calorimetry chambers and coupling these chambers to respirometry equipment to generate "total" or "combined" calorimeter equipment [11, 12, 15–19]. While the adoption of such equipment by the field is very strongly recommended, the fabrication, optimization, validation and use of such equipment is well beyond the current article. Any laboratory planning to develop such equipment is strongly advised to employ a professional electrical and biomedical engineer.

The combined calorimeter system currently operational in our laboratory (Fig. 5), which was custom-fabricated by Colin M.L. Burnett, MD, MS, is thoroughly described in several recent publications [8, 10–12]. Briefly, conditioned air (*see* **Note 1**) is analyzed for flow, temperature, humidity, and pressure to calculate the enthalpy of air entering the chamber. Heat dissipated by the

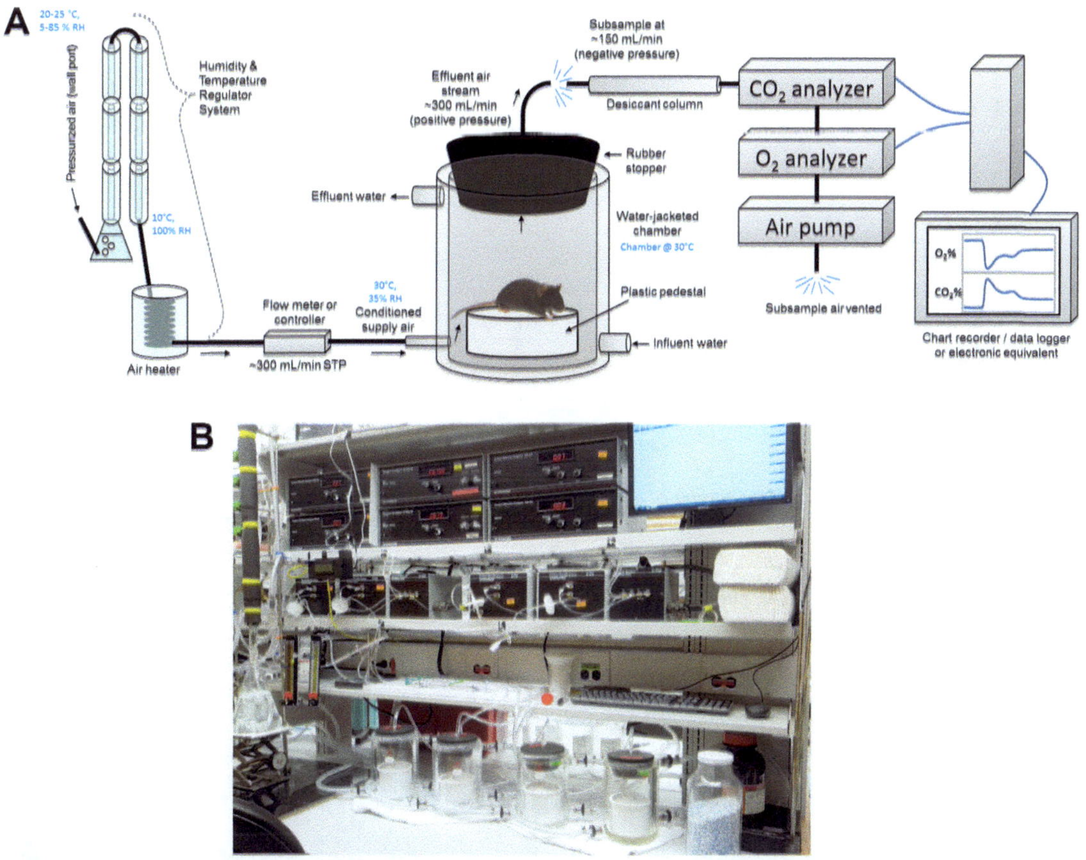

Fig. 4 Illustration of a simple respirometry system. (**a**) A custom-fabricated push-pull continuous respirometry system such as currently in use in our laboratory is illustrated. Pressurized air from a standard laboratory supply line is passed through a humidity and temperature regulator system (explained more fully in Fig. 5), and conditioned air is then passed at ~300 mL/min STP(Standard Temperature and Pressure) into a water-jacketed animal chamber. The animal chamber (a 2-L custom-modified chamber from ACE Glass, Vineland, NJ) is maintained at 30 °C by constant water jacket perfusion via a heated circulating water bath (not shown). Animals are placed into the chamber atop a plastic pedestal (to negate behavioral consequences of placing the animal's thermally sensitive paws upon the heated glass). Air exits the chamber via a port drilled through a tight-fitting rubber stopper that is seated in the mouth of the beaker. Effluent air is then subsampled by a respirometry system, including a desiccant column (anhydrous calcium sulfate/Drierite), a CO_2 analyzer, O_2 analyzer, and finally the air pump that pulls subsampled air through the system. Data from the analyzers is then logged on a chart recorder. (**b**) A photograph of a four-chamber continuous respirometry system currently in use in our laboratory

subject passes through thermopiles that line the interior walls of a water-jacketed chamber, which generates a proportional voltage that is recorded. Flow, temperature, humidity, and pressure are then again measured in effluent air to determine its enthalpy. In addition, core body temperature is recorded using surgically implanted telemeters and a custom miniaturized receiver antenna inside the chamber (Data Sciences International). Effluent air is then subsampled (i.e., flow is less than through the chamber) by a standard respirometry system as above.

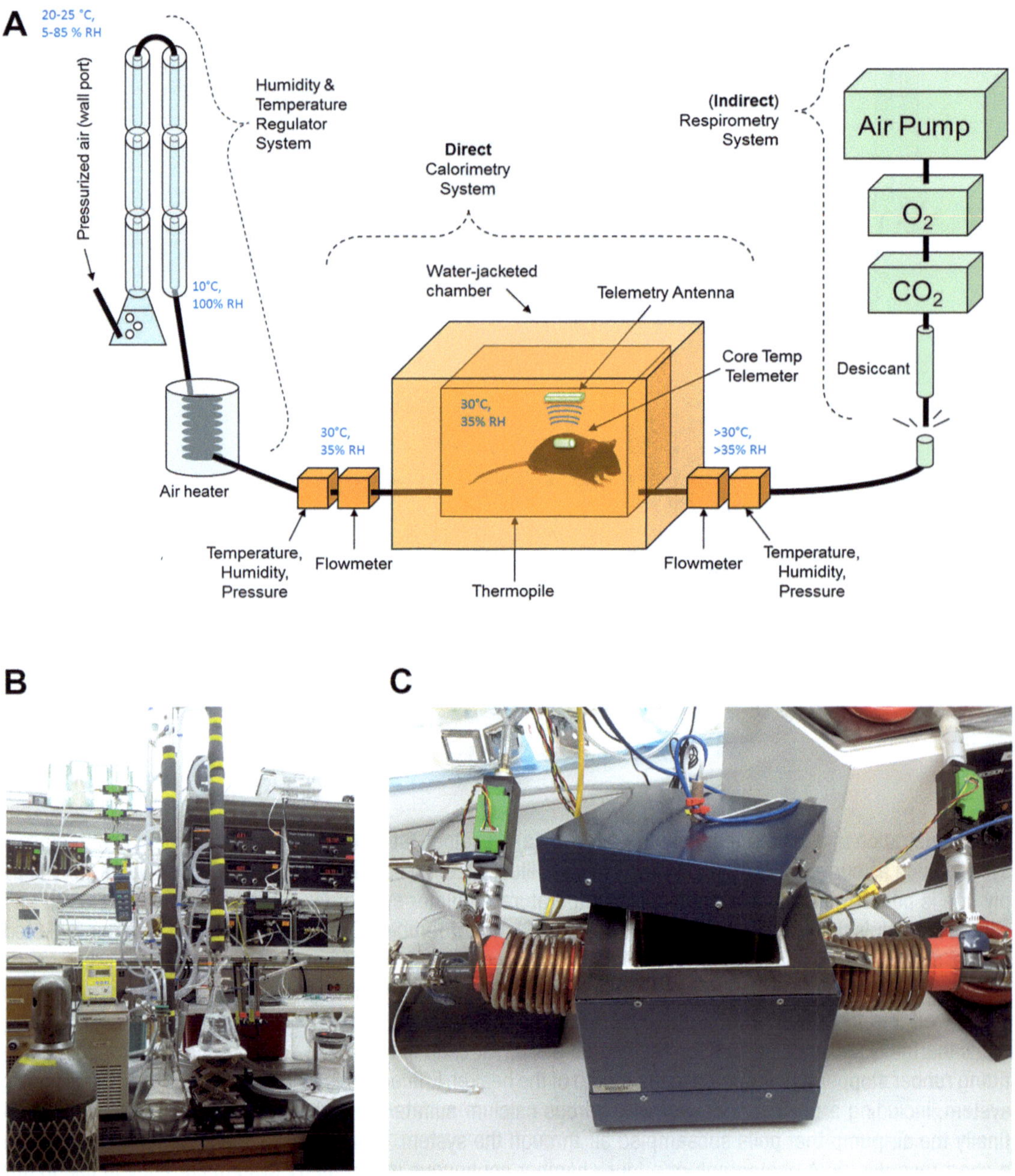

Fig. 5 Illustration of a combined/total calorimeter. (**a**) A custom-fabricated combined calorimeter system is illustrated. First, pressurized air (of uncontrolled humidity and temperature) from a standard laboratory supply line is passed through a humidity and temperature regulator system to supply a constant stream of 30 °C/~35% RH/300 mL/min STP (Standard Temperature and Pressure) air to the chamber (see **Note 1**). Enthalpy of the influent and effluent air streams is calculated by measuring flow, temperature, humidity, and pressure. Heat radiated into the walls of the chamber is determined from the voltage generated by thermopiles that line the interior walls of the water-jacketed chamber. Core body temperature is recorded continuously by radiotelemetric probe. Effluent air is then passed to a respirometry system, as above. (**b**) A photograph of the humidity and temperature regulator system currently in use in our laboratory. (**c**) A photograph of the gradient-layer direct calorimetry chamber currently in use in our laboratory

3 Methods

3.1 Method 1: Metabolic Caging Systems

When an experimental manipulation results in a change in body mass, one simple and superior method to investigate energy balance is to use metabolic caging systems to quantitatively assess food intake. While some investigators prefer to examine food intake behavior in home cages to minimize behavioral interruptions due to neophobia, it is virtually impossible to quantify food intake with high resolution in this setting due to the loss of spilled food dust into bedding materials. Subsequently, food intake in home cages is often reported at higher rates than in metabolic caging systems, but the relative contribution of neophobia-induced anorexia versus food spillage to this difference is impossible to distinguish. Further, mice often consume bedding materials and therefore the caloric content of this additional "food" source is impossible to quantify. (Also, regardless of the caloric value of the bedding material, consumption of that material may also have unexpected consequences on the gut microbiome and thereby energy balance control.) Therefore, we prefer to consistently use metabolic caging systems, but acknowledge the need for explicit training of the animals to minimize the influence of neophobia upon food intake behavior.

Metabolic caging systems (whether commercial or custom-fabricated) typically include a food hopper and fluid delivery system that minimizes spillage of food and drink solutions, and a wire mesh or metal grate floor which allows for efficient and quantitative collection of both urine and fecal samples. Animals should be placed into caging systems for several days to acclimate (*see* **Note 2**) before measurements are performed. Following acclimation, the investigator can measure food intake (mass change of food hopper) and fecal output (mass change of fecal collection tube) daily for several consecutive days. In our laboratory, we typically use at least 48–72 h of acclimation, as our previous studies have demonstrated the instability of body mass, food intake, and feces production when animals are acclimated for shorter periods (Fig. 6).

On a daily basis the investigator should record the animal's body mass, the end masses/volumes for food, water, urine, and feces. Then after resetting the cage, tare masses/volumes for food, water, urine, and feces should again be recorded. Date, time of cage servicing, room temperature and humidity, and operator are also critical information to record for future reference (*see* Table 2 for an example data recording sheet; notably on day 0 there will be no "end" masses/volumes, and on the last day of study there will be no "tare" values). Seasonal variations in ambient temperature and humidity can have major effects on fluid intake behavior and urine production in addition to the rates of evaporation of urine from the collection tube. Evaporation rates can either be

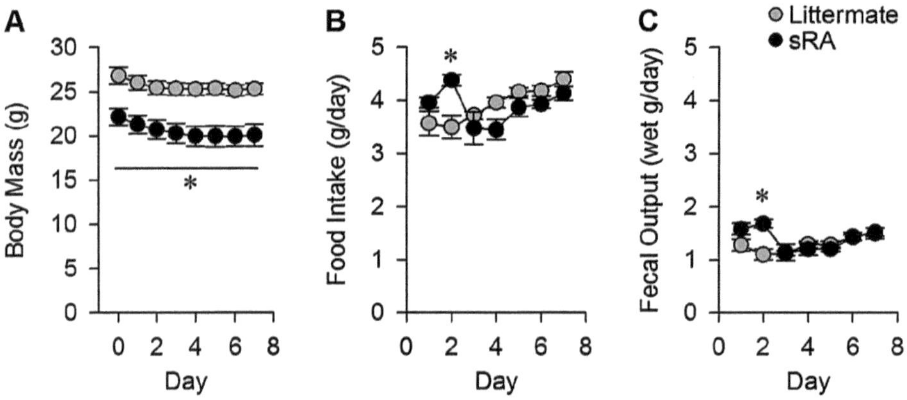

Fig. 6 Example empirical determination of required acclimation time for metabolic caging systems. It is necessary for investigators to empirically determine the requisite acclimation phase for animals in metabolic cages, as acclimation needs can wildly vary based on an array of environmental and animal-specific factors. In this example, double-transgenic "sRA" mice, which exhibit brain-specific hyperactivity of the RAS [3, 6, 34] and their phenotypically normal single- and non-transgenic littermates, were placed into Nalgene-style single-mouse metabolic cages for seven consecutive days and allowed ad libitum access to chow and water. *$P < 0.05$ vs control at indicated timepoint by ANOVA followed by Bonferroni multiple-comparisons procedure

Table 2
Example daily log sheet for metabolic cage studies

Date: ________ Technician: ______________ Room Temperature: ____________
Time: ________ Room Humidity: ____________

Cage #	Animal	Body (g)	End Food (g)	End Water (g or mL)	End Urine (g)	End Feces (g)	Tare Food (g)	Tare Water (g or mL)	Tare Urine (g)	Tare Feces (g)

determined by daily recordings of the mass of water lost from a water-filled tube kept in the animal room, or controlled by placing a small amount of oil in the urine collection tube to "trap" the aqueous urine layer and prevent its evaporation.

Using metabolic caging systems or home cage systems, the investigator can immediately calculate food intake behavior from these data. Using the caloric density of the food (which can be found on the dietary information specifications supplied by the manufacturer), the investigator can then easily calculate the total calories consumed daily by the animal per day (Eq. 1):

$$[\textit{Total calories consumed daily}] = [\textit{Mass of food consumed}] \times [\textit{Caloric density of food}] \quad (1)$$

Using bomb calorimetry methods detailed below, we have determined that the batch-to-batch variability of caloric density of standard chow and high-fat diets from various manufacturers is appreciable. Thus, while the use of a published caloric density is a good starting point, we recommend the routine analysis of the caloric density of individual batches of food for quantitative studies of energy balance using bomb calorimetric methods.

3.2 Method 2: Bomb Calorimetry

By carefully and analytically collecting the feces produced by an animal per unit time, the investigator can then perform a series of tests to qualitatively or quantitatively assess digestive efficiency and caloric absorption. Various methods including fecal acid steatocrit (*see* **Note 3**) or other metabolomics methods can be employed to qualitatively determine whether animals are exhibiting changes in the absorbance rates of fatty acids or other metabolites. To quantitatively assess digestive efficiency and caloric absorption, however, bomb calorimetry is the gold standard. Its less-than-universal use in animal physiology laboratories may stem from the perceived complexity of this physical/analytical chemistry method, or simply from unpalatable but necessary handling and use of feces.

Bomb calorimetry is a methodology that is easily adaptable for use in physiological studies. In essence, this method allows for the determination of the caloric density of a sample material (*see* **Note 4**). After an investigator quantitatively collects feces from study animals via metabolic caging systems (this is extremely difficult to do if the investigator uses home cages), the fecal samples are desiccated. It is critical to calculate the "wet" mass of the feces before desiccation and the "dry" mass after drying. This allows the investigator to calculate the water content of the feces, which may be an endpoint of interest in its own right. Further, it is necessary to know the contribution of "dry" mass to "wet" mass to ultimately calculate digestive efficiency. Desiccated fecal samples are then pressed into pellets, and the pellet mass is carefully determined. It is important to use high-resolution balances for all steps, as significant digit and error propagation issues can become unwieldy after sequential mathematical calculations. Further, the use of an appropriately sized bomb calorimeter is required (*see* **Note 5**).

Fecal samples are then burned to completion in the bomb calorimeter according to the protocol of the individual calorimeter. Typically, the sample is loaded into the bomb along with fuse wire, the bomb is pressurized with pure oxygen, and the bomb is submerged into a known volume of water within a well-insulated container. A very high-resolution thermometer is placed into the water bath, and a water circulating system (such as a paddle wheel or turbine) is submerged and activated. After a suitable thermal stabilization phase, which is subject to ambient conditions and the material composition of various equipment components involved, an electrical current is passed through the bomb's fuse wire and the fecal sample is rapidly combusted to completion. The energy from the combustion is transferred into the water bath, and the resulting temperature change in the water is noted by the thermometer. Having previously calibrated the thermometer system by combusting known masses of pure fuels (e.g., benzoic acid) according to bomb manufacturer's protocol, the user can then calculate the caloric release from the feces. Using that value and the known mass of the fecal pellet that was burned, the investigator can determine the caloric density of the desiccated fecal sample (Eq. 2):

$$[\textit{Caloric density of desiccated feces}] = \frac{[\textit{Energy released from fecal pellet}]}{[\textit{Mass of fecal pellet}]} \tag{2}$$

Once the investigator has determined the caloric density of a desiccated fecal sample, he or she can then calculate the caloric density of the original fecal sample (Eq. 3):

$$[\textit{Caloric density of original sample}] = [\textit{Caloric density of desiccated feces}] \times \left[\frac{[\textit{Mass of dry feces}]}{[\textit{Mass of wet feces}]}\right] \tag{3}$$

Once the caloric density of the original sample has been determined, it is simple to calculate the total daily caloric loss to feces (Eq. 4):

$$[\textit{Total daily caloric loss to feces}] = [\textit{Mass of wet feces}] \times [\textit{Caloric density of original sample}] \tag{4}$$

Next, the investigator can then calculate total daily caloric absorption using the total calories consumed daily (Eq. 1) and the total daily caloric loss to feces (Eq. 4) by simple subtraction (Eq. 5):

$$[\textit{Total daily caloric absorption}] = [\textit{Total calories consumed daily}] - [\textit{Total daily caloric loss to feces}] \tag{5}$$

Total daily caloric absorption is an important value to consider by itself, and the investigator should consider this value as more important to total energy balance than the calories consumed, as ultimately the net energy balance is subject to the balance between calories *absorbed* versus expended.

In addition to total caloric absorption, bomb calorimetry data can be used to calculate digestive efficiency. Digestive efficiency is subject to many different biological mechanisms (mechanics of the gastrointestinal tract, hepatic and pancreatic function, disease states, etc.), but from an energetics perspective these individual mechanisms are only of interest if there is a change in net absorption efficiency. Digestive efficiency simply reflects the fraction of calories absorbed from total calories consumed (Eq. 6):

$$\left[\textit{Digestive efficiency}\right] = \frac{\left[\textit{Total daily caloric absorption}\right]}{\left[\textit{Total calories consumed daily}\right]} \tag{6}$$

Finally, the results of bomb calorimetry can also be used to estimate the relative importance of changes in energy intake versus output in observed changes in body mass. Energy efficiency is a term for the amount of weight gain exhibited by an animal per calorie absorbed over a given period of time (Eq. 7):

$$\left[\textit{Energy efficiency}\right] = \frac{\left[\textit{Change in body mass per unit time}\right]}{\left[\textit{Total calories absorbed per unit time}\right]} \tag{7}$$

Energy efficiency is typically calculated for each of several treatment groups that were matched under baseline conditions (such as before initiation of an experimental diet). For example, after treating two groups of mice with chow or a 45% high-fat diet for 28 days, bomb calorimetry is used to calculate the daily caloric absorption of each animal. For each mouse, energy efficiency is calculated as the weight gained over the 28 days, divided by the total calories absorbed over the 28 days (i.e., the daily absorption × 28 days). A positive deflection is expected in the mice fed the high-fat diet, which would be interpreted as an increase in weight gained for any individual calorie that was absorbed. If the two groups of animals exhibited similar total daily caloric absorption but differences in energy efficiency (e.g., increased with high-fat feeding), then the investigator can conclude that the increased weight gain is due primarily to a reduction in energy expenditure. While the conclusion may seem obvious (e.g., the same calories were absorbed, but one group got larger), the use of this methodology can quantify the effect over multiple time points, which can be used to illustrate the effect of serial treatments or compensatory changes in the animal groups. Further, this methodology allows the investigator to mathematically describe the relationship.

3.3 Method 3: Body Composition

Many laboratories now have access to core facilities that are equipped with sophisticated body composition analyzers. These analyzers come in multiple forms, including dual-energy X-ray absorptiometry (DEXA), nuclear magnetic resonance (NMR), or magnetic resonance imaging (MRI), and each type of technology sports relative benefits and trade-offs. DEXA scanners use X-ray technology to assess lean, fat, and bone masses, and can be used with animals that have metal implants such as implanted minipumps for drug delivery. NMR equipment provides rapid and highly quantitative assessments of lean, fat, and fluid masses, though animals instrumented with metal implants cannot be analyzed in these systems. MRI equipment uses essentially the same methodology as NMR equipment, but is more commonly used for providing images to illustrate body composition rather than simply quantifying body composition. Regardless of method, animals must be either anesthetized or restrained to prevent movement during scanning.

Use of these various methods can provide substantial insight into the effects of treatments upon the subject, as serial, longitudinal measures within an individual can be performed (i.e., baseline and post-treatment values within subject can be easily captured and compared). In addition, data from these types of analyses can be used in conjunction with bomb calorimetry methods to further estimate the contribution of various energy-balance mechanisms to weight gain. For example, lean tissue is generally estimated to have a caloric value of approximately 4 kcal/g, while fat tissue is estimated to have a caloric value of 9 kcal/g. If an animal gains 1 g each of fat and lean tissues, then the investigator is left to account for 13 kcal of increased energy intake/absorption or reduced energy expenditure. If the animal absorbed an additional 10 kcal during the experimental period, then the investigator can roughly estimate that 77% of the extra weight gain was a result of increased caloric absorption (i.e., 10 kcal absorbed, divided by 13 kcal total). By extension, 23% of the extra weight gain (3 of the 13 kcal) likely resulted from reduced energy expenditure.

3.4 Method 4: Physical Activity Methods

One of the mechanisms of energy expenditure available to any animal is spontaneous physical activity, though physical activity for a typical human accounts for only 0–40% of total daily energy expenditure. Various methods have been developed to estimate physical activity in rodents, including systems which quantify distance travelled by photoelectric beam interruption ("beam breaks"), running wheel revolutions, and systems that track proximity between radiotelemetric transmitter implants and stationary receiver antennas.

It is critical to appreciate that *all* methods used to assess the caloric value of physical activity are qualitative. This is due to the wide array of variables that contribute to the energy equivalency of any given physical motion. As a result, the relationship between recorded motion by an animal and the caloric expenditure is of a multivariate form. For example, if a human weighing 70 kg walks up two flights of stairs and a 140 kg human walks up one flight of stairs, it is essentially impossible to calculate how much energy was spent by each subject, and which subject spent more energy. Body composition, exercise conditioning and capacity, muscle efficiency, blood chemistry, feeding status, disease state, height, altitude, ambient temperature, relative humidity, etc. all contribute to the final energy expenditure required by each subject to perform a given motion. Thus, while the above systems can provide critical qualitative insights into circadian rhythms of activity or gross changes in total activity between treatment groups, it remains impossible to calculate energetic equivalents of these motions. Instead, caloric equivalents of any given amount of physical activity are instead typically calculated as the remaining balance between total calories expended (calculated by bomb calorimetry) and calories expended via resting metabolism (Fig. 1).

3.5 Method 5: Aerobic Resting Metabolism Method—Respirometry

Resting metabolic rate (RMR) refers to the energy expended by a body when resting. Typically this refers to an animal that is stationary, awake, and calm but not sleeping, though it is difficult to dissociate "resting" from "sleeping" bouts in rodent species. RMR typically accounts for 60–100% of total energy expenditure, and it is often assessed using respirometric methods due to the availability of high-quality "turn-key" commercial systems [20–22].

Respirometry is one form of "indirect calorimetry," which is a catch-all term describing a collection of methods that estimate metabolic rate by measuring some simple surrogate. For example, "indirect calorimetry" methods also include methods such as isotope dilution. In the case of respirometry, metabolic rate is estimated by measuring respiratory gas exchange. As early as 1928, Lusk published a table of coefficients that related the amount of heat generated per volume oxygen consumed by a combustion reaction of mixtures of fats and carbohydrates [23]. (Popularized by Lusk, this table represented a corrected version of such a table published by Zuntz and Schumburgin [24].) Relatively simple algebraic regressions of these interpolated data lead to an equation to relate rates of oxygen consumption (VO_2) and carbon dioxide production (VCO_2) to heat production [20, 21], which are calculated from the changes in air O_2 and CO_2 content as it passes a live animal (with flow rates measured and corrected for standard temperature and pressure, STP), and many investigators and commercial equipment use this equation to estimate metabolic rate (Eqs. 8–10):

$$[VO_2] = [\Delta O_2\%] \times [STP - corrected\ chamber\ air\ flow] \quad (8)$$

$$[VCO_2] = [\Delta CO_2\%] \times [STP - corrected\ chamber\ air\ flow] \quad (9)$$

$$[Metabolic\ rate] = 3.815 \times [\mathrm{VO}_2] + 1.232 \times [VCO_2] \quad (10)$$

This equation and other similar equations are used by virtually every commercial respirometry-based system, be it a stand-alone system or part of a comprehensive monitoring system.

Respirometry systems can be in many forms [20] (*see* **Note 6**), but many investigators utilize "push-pull" design systems, where air of known composition is "pushed" through an air-tight testing chamber, and effluent air is "pulled" into a subsampling system including a desiccant system to dry the air, then through oxygen and carbon dioxide analyzers. Testing chambers can be maintained at any desired temperature, though assessments of resting metabolism are typically performed at thermoneutral temperatures (e.g., ~30 °C for mice [25]) both to promote "resting" behavior and also to minimize the activation of adaptive thermoregulatary mechanisms (e.g., sweating, licking, panting, shivering, or non-shivering thermogenesis). Various publications have demonstrated that the thermoneutral zone for wild-type C57BL/6 mice is between 26 and 34 °C [25], and therefore most investigators use an ambient/chamber temperature of 30 °C for measurements of RMR. Although perhaps not immediately obvious, it is also important that air flow through such systems must be corrected to Standard Temperature and Pressure (STP) for calculations of metabolic rate, and any custom-fabricated systems must be designed to make this correction. This is necessary because an air flow of, say, 100 mL per hour will not represent the same mass of air if it were at a different temperature or pressure.

RMR measures must be taken while animals are "resting." Investigators can ensure that animals are resting via visual inspection (either in person or via video camera so as to not disturb the resting animal), or more simply by tracking oxygen consumption and carbon dioxide production rates in real-time. Whereas awake and moving animals exhibit highly variable second-to-second rates, resting/sleeping animals will achieve a low, steady plateau of rates that are easily noted in strip charts or electronic equivalents (Fig. 7). Such assessments are only possible if investigators utilize 'continuous' measurement methods; systems with discontinuous measures are not favorable for precise assesments of RMR.

3.6 Method 6: Total Resting Metabolism Method—Direct Calorimetry

It is very important to note that by definition, respirometry is limited to assessments of oxygen-dependent (i.e., aerobic) metabolism. In fact, the use of respirometry relies upon several untested/untestable assumptions which most users ignore [22]. Growing evidence now supports the conclusion that the tacit assumption that anaerobic processes are both negligible and constant is simply wrong. Our group and others have demonstrated that anaerobic processes appear to contribute large, variable fractions of total RMR in various species using "combined calorimetry" or "total calorimetry" methods, which are described below. Walsberg and Hoffman [17, 18] demonstrated that anaerobic processes contribute 0–40% of total RMR in snakes, birds, and kangaroo rats. Pittet et al.

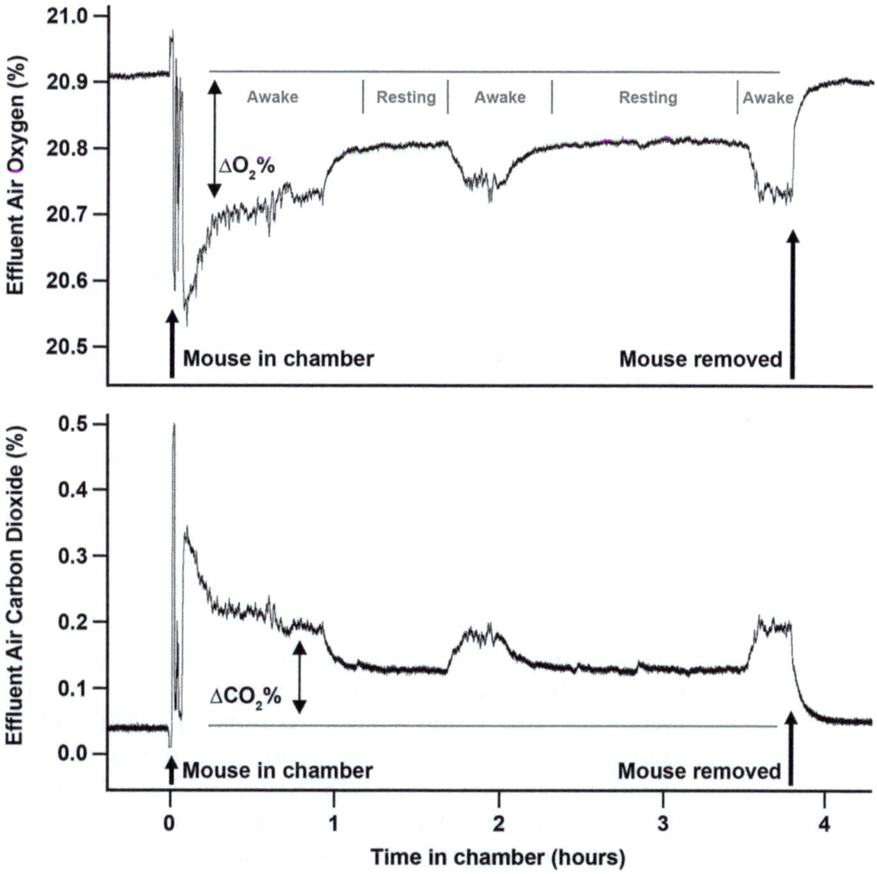

Fig. 7 Example real-time tracing of oxygen consumption and carbon dioxide production by a mouse. A typical recording session of a mouse placed into a (continuous) push-pull respirometer system, with chamber temperature maintained at 30 °C and with air flowing through the chamber at ~300 mL/min STP (Standard Temperature and Pressure). The oxygen consumption rate ($\Delta O_2\%$; reflected in a downward deflection in effluent O_2 content) and carbon dioxide production rate ($\Delta CO_2\%$; reflected in an upward deflection in effluent CO_2 content) can be determined both during bouts of wakefulness (Awake) and during periods of rest (Resting). Periods of rest are easily identified in tracings of effluent air composition as periods of steady, minimum-value plateaus in both $\Delta O_2\%$ and $\Delta CO_2\%$ endpoints. Aerobic metabolic rate is then calculated using Eqs. 8–10, and aerobic RMR simply reflects the output of these calculations when inputting $\Delta O_2\%$ and $\Delta CO_2\%$ values recorded during periods of rest

[15, 16] demonstrated that in lean humans, anaerobic processes contribute nearly 10% of total RMR, while they contribute essentially 0% in obese humans. Van Breukelen's group has also started investigating these processes in desert and aquatic species and in the context of torpor [19, 26–29]. Our group has determined that in laboratory mice, anaerobic RMR contributes around 10% of total RMR, and that this contribution can be suppressed or stimulated by various genetic (e.g., disruption of the angiotensin AT_2 receptor), pharmacological (anesthesia), dietary (high-fat diet), and microbial (risperidone) interventions [8, 10–12]. Thus, there is a major need for the field to adopt methods that can measure the

entire range of processes (aerobic and anaerobic) contributing to RMR [22, 30].

At the chemical level, metabolic rate is the rate of heat production. In steady-state, the rate of heat production by an animal is equivalent to the rate of heat release from the body. Thus, "direct" calorimetry describes methods that measure heat dissipation to calculate metabolic rate. There are currently no commercial sources for direct calorimetry systems for rodents, though several groups have described simplified [19] and more elaborate [11, 22] systems with a range of capabilities and resolutions which can be used for rodents of various sizes. Critically, the investigator must insert/implant a core temperature probe within the animal to either ensure that measurements are performed while the animal is at steady state (indicated by a stable core temperature) or to correct direct calorimeter data for the rate of core temperature change, using complicated equations that depend upon knowledge of the subject's exact body composition [31].

As air must be fluxed through the chamber of a direct calorimeter for the animal to survive, it is straightforward to simply subsample effluent air from a direct calorimeter and perform respirometric measurements to simultaneously measure aerobic metabolic rate. Using such a "combined calorimetry" or "total calorimetry" setup, the investigator can therefore measure total metabolic rate (via direct calorimetry) and aerobic metabolic rate (via respirometry). The difference between these methods, therefore, represents an index of the anaerobic metabolic rate.

Use of a combined calorimeter is essentially identical to the use of a respirometer system. The testing chamber is maintained at an experimentally dictated temperature (typically, thermoneutrality), and the heat production/release within the box and the composition and flow rate of the air fluxing through the box are determined while the animal is in the box and at rest, compared to when the animal is not in the box.

3.7 Method 7: Putting It All Together

As described in the above sections, quantitative evaluations of energy flux in a mouse (or any animal subject) can be efficiently performed using metabolic caging systems and a bomb calorimeter. Additional dissection of the contribution of resting metabolic processes versus physical activity requires far more sophisticated methodologies that are not as commonly (or inexpensively) available. By multiplexing these various methods, however, it is possible to comprehensively evaluate all aspects of energy balance in the animal, and to specifically attribute the relative importance of individual mechanisms to the net changes in body mass (Fig. 1).

Metabolic caging and bomb calorimetry allows the investigator to evaluate caloric intake, absorption, digestive efficiency, and energy efficiency. These methods directly address the contributions of food intake and gastrointestinal function to energy balance. By additionally utilizing body composition analysis methods, investigators can illustrate the relative importance of observed processes to weight gain/loss, compared to what is expected based on changes in body composition. By using quantitative measures of resting metabolism (both aerobic and anaerobic), investigators can then determine the contributions of these processes to total energy balance, and finally by simple subtraction, remaining calories are most easily ascribed to physical activity. Such designations can be complemented/confirmed by qualitative methods of assessing physical activity behavior.

3.8 Method 8: Statistical Considerations

Substantial discussion has raged in the literature over the last decade with regard to the most appropriate way to compare metabolic phenotyping data across subjects of differing body size/body mass/body composition. Of course a larger animal would be expected to eat more food, but documenting a larger food intake in a larger animal does not necessarily indicate that the greater food intake preceded or caused the additional weight gain. As explained in other recent review articles, investigators often normalize data by simply dividing food intake or RMR data by body mass, but from both mathematical and biological perspectives this approach is untenable. Instead, more reasonable solutions to this problem have been proposed to include either simply reporting all of the data (such that the reader is left to draw their own conclusions), or to account for differences in subject body mass using the ANCOVA statistical correction method [32].

As an oversimplification, ANCOVA analyses involve performing linear regressions of metabolic rate (or food intake etc.) versus body mass data pairs. One assumes (and tests) that the slopes of the lines which best describe data from each treatment group are indistinguishable, and then a comparison of the two groups at a single average value for body mass is made (Fig. 8). Statistical packages are available commercially which can perform one-way ANCOVA, or various online calculators for free (*see* **Note** 7). Univariate regression modeling must be used if more than one variable is considered.

4 Notes

1. To control ambient air temperature and humidity for subject animals in calorimeter systems throughout various seasons, a simple system that controls the dew point of the air stream can be used. Water is bubbled through room-temperature water to saturate the air (i.e., 100% relative humidity, RH). This air is then passed through condenser columns maintained by a chilled circulating water bath, thereby "setting" the dew point of the air stream. Warming this air maintains the quantity of

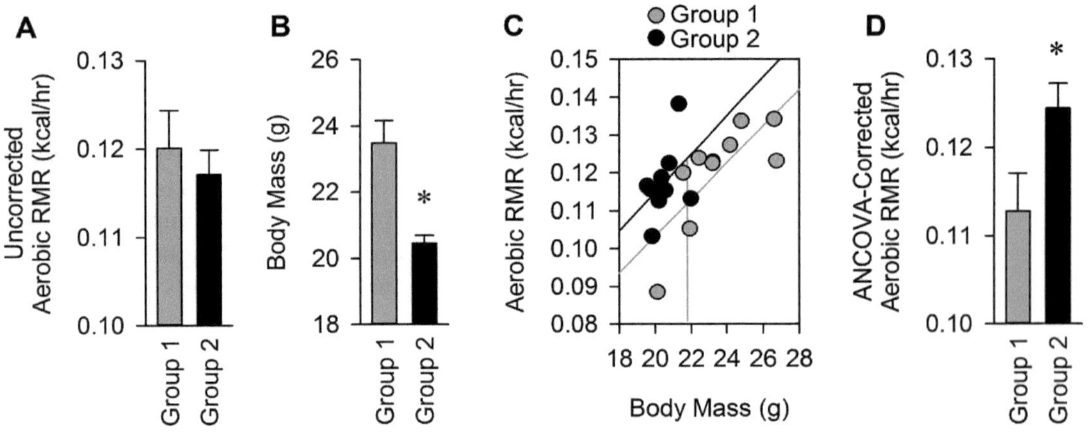

Fig. 8 Illustration of an ANCOVA analysis. (**a**) A simple comparison between two different groups indicates no difference in mean aerobic RMR between groups. (**b**) Comparison of body masses between groups uncovers a difference between groups, necessitating the reanalysis of RMR data to account for body size. (**c**) Metabolic rate values are plotted against the body mass of individual animals within each group, and linear regressions are performed for each group. Tests of homogeneity of regression are performed to confirm that the slopes of lines through each group are indistinguishable. (In this case the lines are indistinguishable, $P = 0.96$, so we can proceed with ANCOVA analyses). (**d**) The interpolated average RMR value (intercept) at the average body mass value for all animals in the study (21.97 g) is then reported as the "ANCOVA-corrected" RMR value

water vapor and reduces the relative humidity. For example, setting the chilled water to 10 °C and warming to 30 °C results in a relative humidity of about 35%. Air flow through this conditioner system can be controlled by modulating the pressure supplied to the system.

2. Because of the neophobia exhibited by rodents, animals will often exhibit abnormal behaviors (and by extension, energy balance) for between 1 and 7 days when placed into metabolic caging systems. Users should empirically determine the acclimation period required for animals in their laboratory, as acclimation needs can differ depending upon the animal strain and the complexity of the caging system. For example, in our laboratory we have repeatedly determined that food and water intake, urine and fecal production all drop over the first 24–96 h when C57BL/6J mice and transgenic mice on this background are placed into Nalgene-style single-mouse sized metabolic cages when housed at 22–25 °C with 30–70% relative humidity, but that intake and output behaviors rebound and achieve a steady plateau after 72–96 h (Fig. 5). In contrast, other laboratories insist upon (based on various levels and types of evidence) waiting up to 7 days of acclimation. We advocate empirical determinations in each laboratory to gain confidence in one's own data.

3. Fatty acid content of feces can be assessed using a very simple biochemical method [8, 33]. Briefly, a known mass of desiccated

feces (e.g., 50 mg) is reconstituted in freshly prepared perchloric acid (200 μL of 1 N). After reconstitution, a small amount of oil-red-O stain (100 μL of 0.5%) is then added to the slurry. The slurry is then loaded in a capillary tube and centrifuged. The layer stained with oil-red-O is then compared (as a percentage) against the entire column in a very similar method to determining the hematocrit of blood. Relative increases or decreases in the contribution of this layer to the total column height reflect decreases or increases, respectively, in fatty acid absorption efficiency. That is, an increase in steatocrit means a decrease in absorption. This qualitative method provides some insight into digestive efficiency, but because it is impossible to determine the caloric value of the fatty acid species present and the value of the other, nonfat components in the feces, this method often serves as only a screening tool or qualitative complement to bomb calorimetric methods. Notably, we have found that when animals are maintained on a chow diet the fatty acid content of the stool is very low and it is difficult to discern changes by this method. In contrast, when animals are maintained on high-fat diets (45 or 60% kcal from fat), changes in fatty acid absorption are easy to detect by this method.

4. It is worth noting that analyses of caloric density using a bomb calorimeter ignore the bioavailability of calories in a sample. For example, a chow-based diet typically contains some amount of undigestible fiber, yet combustion of that food sample in a bomb calorimeter will yield a caloric density including calories from both the digestible and undigestible components. Ultimately if calculations are performed correctly this will not matter as the same undigestible materials are represented in feces, and the same constant value is present in both the consumed and fecal masses. Subtraction of fecal calories from consumed calories (Eq. 5) therefore yields the same value for total daily caloric absorption as this constant is subtracted from itself.
5. A single mouse can be expected to reliably produce ~100 mg of desiccated fecal sample per day, so use of an appropriately sized bomb is necessary. For example, use of a bomb which requires 1 g of sample would necessitate combining fecal samples from 10 individual mice or serially collecting fecal samples from a single mouse for 10 consecutive days. In contrast, a bomb which requires 25–200 mg of sample is suitable to analyze samples from single-day collections from individual mice.
6. Many multichamber metabolic monitoring systems are available commercially through an array of companies including Columbus Instruments, Sable Systems, and TSE. Some of these systems make use of continuous-recording respirometer arrays in which each cage has a dedicated pair of oxygen and carbon dioxide analyzers continuously analyzing effluent air; such systems are analogous to the system described herein.

In contrast, some systems keep costs down by using an automated sampling manifold such that one single analyzer pair is used for a number of chambers. This second, discontinuous-type system samples effluent air from each of several cages for a few seconds every few minutes to provide a survey of metabolic rate values across time. As a result, the addition of more chambers to such a discontinuous system reduces the sampling frequency. As sampling time points are separated, it becomes increasingly difficult to interpret bouts of rest. Therefore for assessments specifically of RMR (as opposed to simply metabolic rate in general), we have had much greater success using a continuous-sampling system as illustrated in Fig. 5.

7. One free source for one-way ANCOVA analyses can be found at: http://vassarstats.net/. It is important to note that if more than one independent variable is in play, that univariate regression methods must be used instead of the simple ANCOVA method.

Acknowledgments

The Grobe laboratory is supported by grants from the NIH (HL134850, HL084207), the American Diabetes Association (1-14-BS-079), the American Heart Association (15SFRN23730000), the University of Iowa Office of the Vice President for Research and Economic Development, the Fraternal Order of Eagles' Diabetes Research Center, and the UIHC Center for Hypertension Research. Some of the equipment illustrated was purchased with support from the Roy J. Carver Trust. Connie C. Grobe, PhD, Jeremy A. Sandgren, John R. Kirby, PhD, and Colin M.L. Burnett, MD, MS provided critical revisions of the chapter text.

References

1. Littlejohn NK, Grobe JL (2015) Opposing tissue-specific roles of angiotensin in the pathogenesis of obesity, and implications for obesity-related hypertension. Am J Physiol Regul Integr Comp Physiol 309(12):R1463–R1473. doi:10.1152/ajpregu.00224.2015
2. Hall KD, Sacks G, Chandramohan D, Chow CC, Wang YC, Gortmaker SL, Swinburn BA (2011) Quantification of the effect of energy imbalance on bodyweight. Lancet 378(9793):826–837. doi:10.1016/s0140-6736(11)60812-x
3. Grobe JL, Grobe CL, Beltz TG, Westphal SG, Morgan DA, Xu D, de Lange WJ, Li H, Sakai K, Thedens DR, Cassis LA, Rahmouni K, Mark AL, Johnson AK, Sigmund CD (2010) The brain renin-angiotensin system controls divergent efferent mechanisms to regulate fluid and energy balance. Cell Metab 12(5):431–442. doi:10.1016/j.cmet.2010.09.011
4. Grobe JL, Dickson ME, Park S, Davis DR, Born EJ, Sigmund CD (2010) Cardiovascular consequences of genetic variation at −6/235 in human angiotensinogen using "humanized" gene-targeted mice. Hypertension 56(5):981–987. doi:10.1161/hypertensionaha.110.157354
5. Grobe JL, Buehrer BA, Hilzendeger AM, Liu X, Davis DR, Xu D, Sigmund CD (2011) Angiotensinergic signaling in the

brain mediates metabolic effects of deoxy-corticosterone (DOCA)-salt in C57 mice. Hypertension 57(3):600–607. doi:10.1161/hypertensionaha.110.165829

6. Littlejohn NK, Siel RB Jr, Ketsawatsomkron P, Pelham CJ, Pearson NA, Hilzendeger AM, Buehrer BA, Weidemann BJ, Li H, Davis DR, Thompson AP, Liu X, Cassell MD, Sigmund CD, Grobe JL (2013) Hypertension in mice with transgenic activation of the brain renin-angiotensin system is vasopressin dependent. Am J Physiol Regul Integr Comp Physiol 304(10):R818–R828. doi:10.1152/ajpregu.00082.2013
7. Grobe JL, Rahmouni K, Liu X, Sigmund CD (2013) Metabolic rate regulation by the renin-angiotensin system: brain vs. body. Pflugers Arch 465(1):167–175. doi:10.1007/s00424-012-1096-9
8. Weidemann BJ, Voong S, Morales-Santiago FI, Kahn MZ, Ni J, Littlejohn NK, Claflin KE, Burnett CM, Pearson NA, Lutter ML, Grobe JL (2015) Dietary sodium suppresses digestive efficiency via the renin-angiotensin system. Sci Rep 5:11123. doi:10.1038/srep11123
9. Fink BD, Herlein JA, Guo DF, Kulkarni C, Weidemann BJ, Yu L, Grobe JL, Rahmouni K, Kerns RJ, Sivitz WI (2014) A mitochondrial-targeted coenzyme q analog prevents weight gain and ameliorates hepatic dysfunction in high-fat-fed mice. J Pharmacol Exp Ther 351(3):699–708. doi:10.1124/jpet.114.219329
10. Bahr SM, Weidemann BJ, Castro AN, Walsh JW, deLeon O, Burnett CM, Pearson NA, Murry DJ, Grobe JL, Kirby JR (2015) Risperidone-induced weight gain is mediated through shifts in the gut microbiome and suppression of energy expenditure. EBioMedicine 2(11):1725–1734. doi:10.1016/j.ebiom.2015.10.018
11. Burnett CM, Grobe JL (2013) Direct calorimetry identifies deficiencies in respirometry for the determination of resting metabolic rate in C57Bl/6 and FVB mice. Am J Phys Endocrinol Metab 305(7):E916–E924. doi:10.1152/ajpendo.00387.2013
12. Burnett CM, Grobe JL (2014) Dietary effects on resting metabolic rate in C57BL/6 mice are differentially detected by indirect (O2/CO2 respirometry) and direct calorimetry. Mol Metab 3(4):460–464. doi:10.1016/j.molmet.2014.03.003
13. Xu D, Borges GR, Davis DR, Agassandian K, Sequeira Lopez ML, Gomez RA, Cassell MD, Grobe JL, Sigmund CD (2011) Neuron- or glial-specific ablation of secreted renin does not affect renal renin, baseline arterial pressure, or metabolism. Physiol Genomics 43(6):286–294. doi:10.1152/physiolgenomics.00208.2010
14. Hilzendeger AM, Cassell MD, Davis DR, Stauss HM, Mark AL, Grobe JL, Sigmund CD (2013) Angiotensin type 1a receptors in the subfornical organ are required for deoxy-corticosterone acetate-salt hypertension. Hypertension 61(3):716–722. doi:10.1161/hypertensionaha.111.00356
15. Pittet P, Gygax PH, Jequier E (1974) Thermic effect of glucose and amino acids in man studied by direct and indirect calorimetry. Br J Nutr 31(3):343–349
16. Pittet P, Chappuis P, Acheson K, De Techtermann F, Jequier E (1976) Thermic effect of glucose in obese subjects studied by direct and indirect calorimetry. Br J Nutr 35(2):281–292
17. Walsberg GE, Hoffman TC (2005) Direct calorimetry reveals large errors in respirometric estimates of energy expenditure. J Exp Biol 208(Pt 6):1035–1043. doi:10.1242/jeb.01477
18. Walsberg GE, Hoffman TC (2006) Using direct calorimetry to test the accuracy of indirect calorimetry in an ectotherm. Physiol Biochem Zool 79(4):830–835. doi:10.1086/505514
19. Burger M, van Breukelen F (2013) Construction of a low cost and highly sensitive direct heat calorimeter suitable for estimating metabolic rate in small animals. J Therm Biol 38(8):508–512. doi:10.1016/j.jtherbio.2013.09.002
20. Lighton JRB (2008) Measuring metabolic rates: a manual for scientists. Oxford University Press, Oxford. https://global.oup.com/academic/product/measuring-metabolic-rates-9780195310610?cc=us&lang=en& DOI: 10.1093/acprof:oso/9780195310610.001.0001
21. McLean JA, Tobin G (2007) Animal and human calorimetry. Cambridge University Press, Cambridge
22. Kaiyala KJ, Ramsay DS (2011) Direct animal calorimetry, the underused gold standard for quantifying the fire of life. Comp Biochem Physiol A Mol Integr Physiol 158(3):252–264. doi:10.1016/j.cbpa.2010.04.013
23. Lusk G (1928) The elements of the science of nutrition, 4th edn. W.B. Saunders Company, Philadelphia, PA
24. Zuntz, N. 1847-1920. (1901). Studien zu einer Physiologie des Marsches: von Zuntz und Schumburg. Berlin: Hirschwald. pp 1847–1920
25. Maloney SK, Fuller A, Mitchell D, Gordon C, Overton JM (2014) Translating animal model research: does it matter that our rodents are cold? Physiology (Bethesda) 29(6):413–420. doi:10.1152/physiol.00029.2014

26. Utz JC, Velickovska V, Shmereva A, van Breukelen F (2007) Temporal and temperature effects on the maximum rate of rewarming from hibernation. J Therm Biol 32(5):276–281. doi:10.1016/j.jtherbio.2007.02.002
27. Utz JC, van Breukelen F (2013) Prematurely induced arousal from hibernation alters key aspects of warming in golden-mantled ground squirrels, *Callospermophilus lateralis*. J Therm Biol 38(8):570–575. doi:10.1016/j.jtherbio.2013.10.001
28. Heuton M, Ayala L, Burg C, Dayton K, McKenna K, Morante A, Puentedura G, Urbina N, Hillyard S, Steinberg S, van Breukelen F (2015) Paradoxical anaerobism in desert pupfish. J Exp Biol 218(Pt 23):3739–3745. doi:10.1242/jeb.130633
29. van Breukelen F, Martin SL (2015) The hibernation continuum: physiological and molecular aspects of metabolic plasticity in mammals. Physiology (Bethesda) 30(4):273–281. doi:10.1152/physiol.00010.2015
30. Kaiyala KJ (2014) What does indirect calorimetry really tell us? Mol Metab 3(4):340–341. doi:10.1016/j.molmet.2014.03.005
31. Blaxter K (1989) Energy metabolism in animals and man. Cambridge University Press, Cambridge
32. Tschop MH, Speakman JR, Arch JR, Auwerx J, Bruning JC, Chan L, Eckel RH, Farese RV Jr, Galgani JE, Hambly C, Herman MA, Horvath TL, Kahn BB, Kozma SC, Maratos-Flier E, Muller TD, Munzberg H, Pfluger PT, Plum L, Reitman ML, Rahmouni K, Shulman GI, Thomas G, Kahn CR, Ravussin E (2012) A guide to analysis of mouse energy metabolism. Nat Methods 9(1):57–63. doi:10.1038/nmeth.1806
33. Takahashi N, Li F, Hua K, Deng J, Wang CH, Bowers RR, Bartness TJ, Kim HS, Harp JB (2007) Increased energy expenditure, dietary fat wasting, and resistance to diet-induced obesity in mice lacking renin. Cell Metab 6(6):506–512. doi:10.1016/j.cmet.2007.10.011
34. Sakai K, Agassandian K, Morimoto S, Sinnayah P, Cassell MD, Davisson RL, Sigmund CD (2007) Local production of angiotensin II in the subfornical organ causes elevated drinking. J Clin Invest 117(4):1088–1095. doi:10.1172/jci31242

Chapter 11

In Vitro Assays to Determine Smooth Muscle Cell Hypertrophy, Protein Content, and Fibrosis

Katherine J. Elliott and Satoru Eguchi

Abstract

This chapter provides information on how to culture primary rat vascular smooth muscle cells and how to induce cellular changes similar to those associated with angiotensin II activation in vivo. We describe how to assess the cellular changes by determining cell size with an automated coulter cell counter to measure cell volume. In addition, we describe a method to assess total protein content. Finally, we describe a standard technique to quantify angiotensin II-induced pro-fibrotic response using the Chondrex Sirius Red Total Collagen Detection Kit.

Key words Explant primary cell culture, VSMC, Angiotensin II, Hypertrophy, Fibrosis

1 Introduction

This chapter provides the reader with instructions on how to culture rat vascular smooth muscle cells (VSMC) and how to stimulate them with angiotensin II (Ang II) to induce a cellular hypertrophic phenotype similar to that observed in hypertension. Primary cell cultures, derived from dissection of living tissue and maintained in culture for a limited number of passages, provide a great tool to simplify the study of complex biologic disease processes such as hypertension and associated organ damage. This explant method allows VSMC to migrate from the rat aorta onto a cell culture dish forming a monolayer of adherent cells, which will continue to proliferate for up to 12 passages [1, 2]. These cells can be stimulated by various agonists, and downstream physiologic changes such as hypertrophy [3] and pro-fibrotic responses can be assessed [4–6]. This is a relatively easy to use and inexpensive model for the study of pathophysiological changes occurring in cells with Ang II stimulation. We do not include descriptions of cell proliferation assays because Ang II does not stimulate proliferation in VSMC that have been isolated via the explant method [3].

Sean E. Thatcher (ed.), *The Renin-Angiotensin-Aldosterone System: Methods and Protocols*, Methods in Molecular Biology, vol. 1614, DOI 10.1007/978-1-4939-7030-8_11,

2 Materials

2.1 Explant Method for VSMC Primary Cell Culture

This technique was adapted from Ross et al. [2, 4].

1. 12-week-old Male Sprague Dawley rat.
2. Surgical scissors and tweezers.
3. VSMC Growth Medium: Dulbecco's Modification of Eagle's Medium (DMEM: 4.5 g/l glucose), 10% (v/v) fetal bovine serum, 1% (v/v) penicillin–streptomycin solution.
4. 0.05% trypsin–0.53 mM EDTA.
5. Hank's buffered saline solution (HBSS).
6. Tissue culture incubator (37 °C, 5% CO_2).

2.2 VSMC Stimulation with Angiotensin II

1. Serum-free medium: DMEM, 1% (v/v) penicillin–streptomycin solution.
2. Angiotensin II.

2.3 VSMC Hypertrophy

1. 0.5% SDS.
2. BCA protein assay kit.
3. Plate reader or spectrophotometer.
4. Trypsin.
5. HBSS.
6. Phosphate buffered saline (PBS).
7. Scepter Handheld Automated Cell Counter with 60 μm Scepter sensors.

2.4 VSMC Fibrosis

1. Chondrex Sirius Red Total Collagen Detection Kit (Catalog #9062).
2. Pepsin (porcine gastric mucosa) 1 mg/ml in 0.05 M acetic acid.
3. Cell scraper.

3 Methods

3.1 Explant Method for VSMC Primary Cell Culture

1. Excise the thoracic aorta of a male Sprague Dawley rat (200–250 g) directly below the left subclavian artery and above the diaphragm. Rapidly immerse in VSMC growth medium. Carefully remove connective tissue and adherent fat.
2. Longitudinally cut open isolated aorta and remove endothelium by gently rubbing the intimal surface with a dull scissor (*see* **Note 1**).
3. Cut denuded aorta into approximately 3-mm square sections. Place intimal side down into 6-well culture dish.

4. Gently add enough VSMC Growth medium to minimally cover the tissue, about 100–200 μl, without disturbing the orientation of the explant and without causing the explant to float. Culture at 37 °C in 5% CO_2 incubator. Avoid handling plate for 48 h after explant so as not to dislodge the tissue (*see* **Note 2**).
5. Beginning on day 2, monitor plate for cell growth and add additional medium as necessary to avoid drying of tissue and migrating cells.
6. Allow VSMCs to grow out from the explant for 7–14 days (Fig. 1).
7. Once cells have established a monolayer culture surrounding the tissue, remove tissue and culture medium, rinse with PBS, and harvest cells by trypsinization. Transfer collected cells to a 100-mm flask.
8. The cells obtained by this method should express the "hills and valleys" growth characteristics of cultured VSMC. The expression of myosin light chain and smooth muscle α-actin can be confirmed by immunocytochemistry.

Cells should be maintained by passaging once a week with trypsin and seeding in 75-cm^2 flasks at approximately $0.5–1.0 \times 10^6$ cells. Cells should not be passaged more than 12 times and should never be frozen for storage (*see* **Note 3**).

3.2 VSMC Stimulation with Angiotensin II (Ang II)

1. Seed VSMC on 6- or 12-well plate and allow to grow to confluence.
2. Serum-starve confluent VSMC for 48–72 h by removing growth medium, wash cells with serum-free medium and incubate with 1 ml of serum-free medium (Table 1).
3. On the day of stimulation, replace serum-free medium and let cells rest for 1 h in incubator.

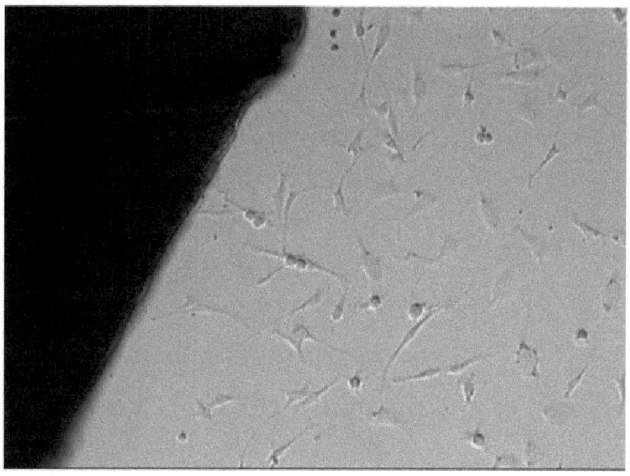

Fig. 1 VSMC explant culture, 14 days post-explant

Table 1
Guidelines for seeding VSMC in different sized culture wells

Plate size	Seeding density	Days to confluence	Approximate cell number at confluence
6 well	2×10^5	3	5×10^5
12 well	1×10^5	2	2×10^5
24 well	5×10^4	2	1×10^5
96 well	3×10^3	2	2×10^4

4. Add Ang II at a final concentration of 100 nM. Stimulation time should be 24–72 h in order to observe maximal hypertrophy (*see* **Note 4**).

3.3 VSMC Hypertrophy

Cellular protein measurement/BCA Assay

1. Seed VSMC in 24-well plates with $n = 6$ for each sample. Allow cells to grow to confluence. Stimulate cells with agonist as needed for experiment.
2. Wash cells with PBS.
3. Lyse cells in well using 150 μl of 0.5% SDS. Let stand for 5 min.
4. Add 25 μl cell lysate to 96-well plate and add BCA reagent (mix according to package instructions).
5. Prepare standard curve.
6. Measure protein amount with plate reader at 570 nm (Fig. 2).

Cell volume

1. Seed VSMC in 6-well plates with $n = 3$ for each sample. Allow cells to grow to confluence. Stimulate cells with agonist as needed for experiment.
2. Wash cells twice with HBSS.
3. Remove cells from plate with 250 μl of trypsin. Add 750 μl PBS to each well. Pipet up and down to disrupt cell clumps and transfer cell suspension to 1.5 ml tube.
4. Centrifuge tube at 250 × *g* for 3 min. Carefully remove supernatant.
5. Resuspend cells in 100 μl PBS.
6. Dilute cells in PBS to an acceptable concentration (*see* **Note 5**). The Coulter Cell Counter can measure 10,000–500,000 cells/ml (Fig. 3).

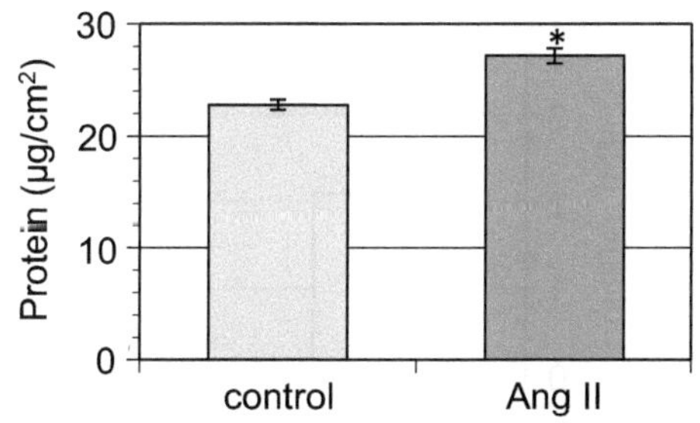

Fig. 2 Representative graph illustrating hypertrophy via increased total protein in AngII-stimulated VSMC cultures. VSMC were plated in a 12-well dish, grown to confluence, and serum-starved for 48 h prior to AngII-stimulation for 72 h. $*p < 0.05$ ($n = 6$)

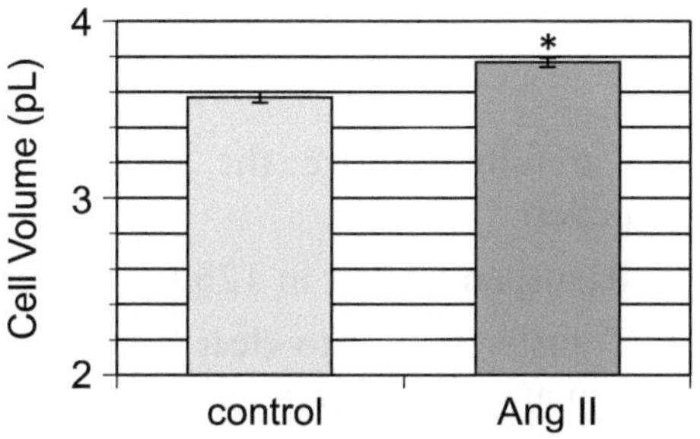

Fig. 3 Representative graph illustrating hypertrophy via increased cell volume in AngII-stimulated VSMC cultures. VSMC were plated in a 6-well dish, grown to confluence, and serum-starved for 48 h prior to AngII-stimulation for 72 h. $*p < 0.05$ ($n = 6$)

3.4 VSMC Fibrosis

1. Seed VSMC in 12-well plates with $n = 6$ for each sample. Allow cells to grow to confluence. Stimulate cells with agonist as needed for experiment.
2. Wash cells with cold, distilled water.
3. Add 50 µl of 0.05 M acetic acid. Scrape cell layer with cell scraper and transfer to 1.5 ml Eppendorf tubes.
4. Rinse wells with 50 µl 0.05 M acetic acid and transfer to respective Eppendorf tube.
5. Add 10 µl of pepsin solution sample. Incubate at 4 °C for 48 h on a rotator.
6. Add 250 µl of Sirius Red solution to each tube. Vortex and incubate at room temperature for 20 min followed by centrifugation at 10,000 × *g* for 3 min.
7. Carefully remove the supernatant without disturbing the pellet (*see* **Note 6**).
8. Resuspend pellet in 250 µl of Washing Solution. Centrifuge at 10,000 × *g* for 3 min.

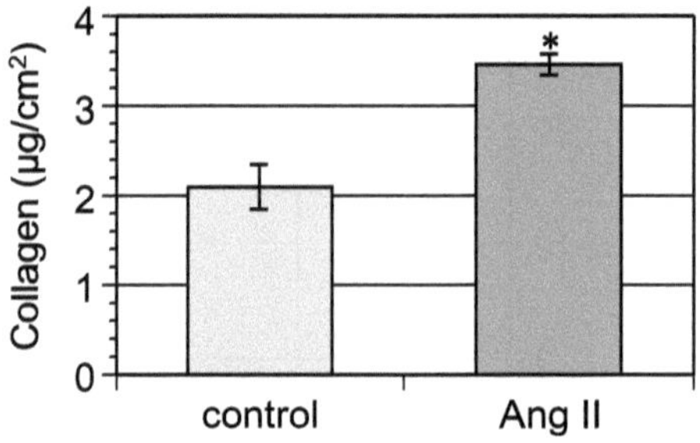

Fig. 4 Representative graph illustrating fibrosis via increased collagen levels in AngII-stimulated VSMC cultures. VSMC were plated in a 12-well dish, grown to confluence, and serum-starved for 48 h prior to AngII-stimulation for 48 h. $*p < 0.05$ ($n = 6$)

9. Carefully remove the supernatant without disturbing the pellet.
10. Resuspend pellet in 125 μl Extraction Buffer. Vortex briefly.
11. Transfer 100 μl to clear bottom 96-well plate. Measure collagen amount with plate reader at 570 nm (Fig. 4).
12. Prepare standard curve by same method as samples.

4 Notes

1. From this point, an aseptic technique should be performed inside a tissue culture hood.
2. It is important that the tissue remain fixed with the intimal side facing the plate so that the VSMC can attach and migrate onto the floor of the culture well.
3. Seeding density may need to be adjusted according to how quickly the cells are growing. Unpublished observations indicate that freezing VSMC results in a phenotypic change in their response to Ang II stimulation. Freezing the cells will select for a subset of highly proliferative cells.
4. Ang II is prepared in H_2O and stored in small aliquots at −20 °C. Store 5 μl of 1 mM (10,000×) stock solution in small tube. Dilute 1:10 with water (1000×) solution.
5. In our hands, we have found that a dilution of 1:100 works well.
6. Using a p200 pipet to remove the supernatant is safer than trying to pour off or aspirate the supernatant because the pellet can be small and soft.

References

1. Geisterfer AA, Peach MJ, Owens GK (1988) Angiotensin ii induces hypertrophy, not hyperplasia, of cultured rat aortic smooth muscle cells. Circ Res 62:749–756
2. Eguchi S, Hirata Y, Imai T, Kanno K, Marumo F (1994) Phenotypic change of endothelin receptor subtype in cultured rat vascular smooth muscle cells. Endocrinology 134:222–228
3. Ohtsu H, Higuchi S, Shirai H, Eguchi K, Suzuki H, Hinoki A, Brailoiu E, Eckhart AD, Frank GD, Eguchi S (2008) Central role of gq in the hypertrophic signal transduction of angiotensin ii in vascular smooth muscle cells. Endocrinology 149:3569–3575
4. Ross R (1971) The smooth muscle cell. II. Growth of smooth muscle in culture and formation of elastic fibers. J Cell Biol 50:172–186
5. Nakashima H, Frank GD, Shirai H, Hinoki A, Higuchi S, Ohtsu H, Eguchi K, Sanjay A, Reyland ME, Dempsey PJ, Inagami T, Eguchi S (2008) Novel role of protein kinase C delta Tyr311 phosphorylation in vascular smooth muscle cell hypertrophy by angiotensin II. Hypertension 51:232–238
6. Takayanagi T, Forrester FJ, Kawai T, Obama T, Tsuji T, Elliott KJ, Nuti E, Rossello A, Kwok HF, Scalia R, Rizzo V, Eguchi S (2016) Vascular ADAM17 as a novel therapeutic target in mediating cardiovascular hypertrophy and perivascular fibrosis induced by angiotensin II. Hypertension 68(4):949–955

Chapter 12

A New Mouse Model for Introduction of Aortic Aneurysm by Implantation of Deoxycorticosterone Acetate Pellets or Aldosterone Infusion in the Presence of High Salt

Shu Liu, Ming C. Gong, and Zhenheng Guo

Abstract

Dysfunction of the renin-angiotensin-aldosterone system (RAAS) has been implicated in the etiologies of many cardiovascular diseases, including aortic aneurysm. In particular, the infusion of angiotensin II (Ang II) in the apolipoprotein E-deficient mice (apoE−/−) and low density lipoprotein receptor knockout mice (LDLR−/−) to induce aortic aneurysm has been extensively used in the field. In contrast, whether aldosterone (Aldo), an essential component of RAAS and a downstream effector of Ang II, is involved in aortic aneurysm is largely unknown. Here, we describe a new animal model for induction of aortic aneurysm in mice in which administration of deoxycorticosterone acetate (DOCA) and high salt or aldosterone and high salt, but not DOCA or high salt alone, to C57BL/6 male mice can potently induce aortic aneurysm formation and rupture in an age-dependent manner. This new aortic aneurysm mouse model is different from Ang II infusion mouse model and exhibits several unique features that mimic human aortic aneurysm.

Key words Aneurysm, Angiotensin, Aldosterone, Sodium, Deoxycorticosterone

1 Introduction

Aortic aneurysm is defined as a permanent localized dilation of the aorta with at least 50% increase in diameter compared with the normal aortic diameter [1]. Aortic aneurysm can be classified according to location as thoracic aortic aneurysm (TAA) and abdominal aortic aneurysm (AAA). AAA is the most common form of aortic aneurysm [2], affecting 4–8% of men and 0.5–1.5% of women over the age of 60, and accounting for nearly 2% of all deaths in Western countries [3, 4]. Due to an incomplete understanding of molecular mechanisms and pathophysiologic processes, no drug has been approved for treatment of progression and rupture of either type of this devastating disease.

Several aortic aneurysm animal models have been developed over the last few decades [3]. Among them, the infusion of Ang II in the apoE−/− and LDLR−/− mice to induce aortic aneurysm is

Sean E. Thatcher (ed.), *The Renin-Angiotensin-Aldosterone System: Methods and Protocols*, Methods in Molecular Biology, vol. 1614,
DOI 10.1007/978-1-4939-7030-8_12, © Springer Science+Business Media LLC 2017

mostly used [5–10]. In contrast to overwhelming evidence for a significant role of Ang II in aortic aneurysm [3, 4], little is known whether aldosterone (Aldo), an essential component of the RAAS and a downstream effector of Ang II, is involved in aortic aneurysm. We recently developed a new mouse model for aortic aneurysm in which administration of mineralocorticoid receptor (MR) agonists, DOCA or Aldo plus high salt, but not DOCA or high salt alone, to C57BL/6 male mice can potently induce aortic aneurysm formation and rupture in an age-dependent manner [11]. Both AAA and TAA could be observed in aortic aneurysm induced by DOCA or Aldo plus high salt but AAA was mostly evident and TAA appeared as a result of AAA progression [11]. Compared to the Ang II infusion mouse model, DOCA or Aldo plus high salt mouse model described here exhibits several unique features, including: (1) using WT mice, not the apoE−/− and LDLR−/− mice, avoids the potential confounding effects of hyperlipidemia on AAA; (2) infusing the mice with 200 μg/kg/day Aldo (which is sufficient to induce aortic aneurysm) increases plasma Aldo concentration to 10.58 nM [11], which is below that in patients with heart failure [12, 13]. This indicates that the Aldo-salt model is a pathological model that mimics human diseases rather than a pharmacological model that would cause concerns due to use of high doses of the molecule of interest that never happens in human patients; (3) using 10-month-old mice rather than 10-week-old mice (mostly used by the Ang II and other models) resembles human AAA; (4) dependent on MR but independent on Ang II receptor since eplerenone (a specific MR antagonist), but neither losartan (an Ang II receptor blocker) nor enalapril (an angiotensin converting enzyme inhibitor), effectively blocks Aldo and high salt induced AAA [11]; (5) requiring high salt for DOCA or Aldo to induce aortic aneurysm implies that high salt intake may be a new risk factor for aortic aneurysm.

Given that there are no effective drug therapies for aortic aneurysms, DOCA or Aldo plus high salt mouse model described here not only provides a new aortic aneurysm animal model induced by MR agonist and high salt but reveals a previously unrecognized but potentially significant role of aldosterone and high salt in the pathogenesis of aortic aneurysm.

2 Materials

2.1 Animals, Chemicals, Osmotic Minipumps, and Equipment

Ten-month-old male C57BL/6J or C57BL/6N mice are either purchased from colonies of the National Institute on Aging (Charles River, Raleigh, NC, USA; *see* **Note 1**) or from the Jackson Laboratory (Bar Harbor, ME, USA) as retired breeders (*see* **Note 2**). All protocols are approved by the University of Kentucky Institutional Animal Care and Use Committee (IACUC).

1. DOCA pellets (50 mg, 21-day release, *see* **Note 3**).
2. d-Aldosterone (Aldo)(200 μg/kg/day) and solubilized in 50% dimethyl sulfoxide (DMSO, *see* **Note 4**).
3. Isoflurane and/or ketamine–xylazine.
4. Depilatory cream (e.g., Nair®).
5. Surgical instruments (forceps, hemostat, scissors, and auto-stapler).
6. Osmotic minipumps (Alzet model 2004, 28-day release, *see* **Note 4**).
7. High salt: 0.9% NaCl plus 0.2% KCl.
8. Noninvasive CODA® tail-cuff blood pressure system.
9. High-resolution ultrasound imaging Vevo® 2100 System.
10. Nikon SMZ800 Stereo microscope with digital imaging software NIS-Elements (Nikon Instruments Inc.).

3 Methods

3.1 Preparation for DOCA Pellet Implantation, Surgical Procedure, and Administration of High Salt in Drinking Water

Mice are anesthetized by intraperitoneal injection of ketamine (100 mg/kg) and xylazine (10 mg/kg; *see* **Note 5**).

1. Shave mice and then use a depilatory cream to remove all hair from the surgical site.
2. Lift the skin on the lateral side of the neck of the animal.
3. Make an incision equal in diameter to that of the pellet.
4. Make a pocket horizontally with a pair of forceps about 2 cm beyond the incision site.
5. Put DOCA pellets into the pocket with forceps.
6. Close the wound with wound clips. Two clips will normally suffice.
7. Mice are given with the drinking water containing high salt (0.9% NaCl plus 0.2% KCl) immediately after DOCA implantation for 3 weeks (*see* **Note 6**).

3.2 Measurement of Mouse Blood Pressure by Tail-Cuff, Quantification of Abdominal Aortic Dilation by Ultrasound Imaging, and External Abdominal Aortic Diameter Measurements

Systolic blood pressure (SBP) is measured using a noninvasive tail cuff system. Measurements are performed for 5 consecutive days for determination of weekly average measures as described [11]. SBP is measured 1 week before DOCA administration (basal) and again during the third week after DOCA and salt administration (*see* **Note 7**).

A high-resolution ultrasound imaging system is used to image and quantify intraluminal diameters of mouse suprarenal abdominal aortas [11, 14]. Briefly, mice are anesthetized by inhalation of isoflurane mixed with O2 (3–5% isoflurane/97% O_2) and maintained by inhalation of isoflurane mixed with O_2 (1–3% isoflurane/97% O_2) throughout the procedure. Mouse hair is removed from the abdomen by using a depilatory cream. Mice are laid supine on a heated table at 37 °C. Warmed ultrasound transmission gel is placed on the abdomen. Cine loops of 300 frames are acquired throughout the renal region of the abdominal aorta and used to determine the maximal diameters of the abdominal aorta in the suprarenal region. "Portal Triad" (hepatic artery, hepatic vein, and bile duct) is used as anatomic marker to assess transverse images in the same location in each mouse. Ultrasound images are acquired 1 week before and weekly after DOCA or aldosterone administration, and are independently validated by two different operators.

Based on the definition of human aortic aneurysm, AAA and TAA are defined in the current study as having at least a 50% increase in maximal intraluminal and external diameters compared with the same region of aorta in control mice. Maximal intraluminal diameters of suprarenal abdominal aortas are quantified in vivo by ultrasound imaging as described above. Maximal external diameters of thoracic and abdominal aortas are photographed by Nikon SMZ800 stereomicroscope with digital imaging software NIS-Elements and then quantified as described [6, 11, 14].

3.3 Induction of Aortic Aneurysm by Aldo and High Salt

1. Weigh study mice before calculating the amount of Aldo needed for infusion (*see* **Note 8**).
2. Use the template (Table 1) to calculate the Aldo mass needed for the experiment. Calculate a 300 μL total volume of Aldo solution or each mouse since each pump requires approximately 250 μL.
3. Weigh the calculated Aldo mass (4.6 mg as shown in Table 1) into a sterile plastic tube.
4. Add a half calculated volume of DMSO (1800 μL) into the plastic tube containing the Aldo, cap, and mix thoroughly by vortexing until the solution is clear.
5. Add another half calculated volume of room temperature dH_2O (1800 μL) into the Aldo solution, cap, and mix thoroughly by vortexing until the solution is clear.
6. Label mouse numbers #1, #2, #3, etc. on individual sterile plastic tubes with caps. Prepare Aldo solution for each mouse based on body weight as shown in Table 1.

3.4 Osmotic Pump Filling

1. Weigh each pump (including both the main pump body and flow moderator). This weight, termed "Pump Weight empty" in the template (Table 1), will be used to calculate the filled ratio.
2. Attach the pump filling needle to a 1 mL sterile syringe and carefully fill the syringe with Aldo solution from the appropriately

Table 1
Calculation of Aldo for osmotic pump infusion

Experimental characteristics				
Dose required	139 ng/kg/min = 200 μg/kg/day			
Start body weight (largest mouse)	37.9 g			
Total estimated body weight gain	1.0 g			
Pump rate	0.25 μL/h			
Start body weight (largest mouse)	37.9 g			
Number of mice	12			
Calculations (see **Note 8**)				
Does per hour for animals	320.256 ng[a]			
Concentration needed	1281. 024 ng/μL[b]			
For 300 μl solution	0.384 mg/300 μL[c]			
Solution needed				
Total aldosterone (mg)	4.608 mg[d]			
Dissolved in (DMSO + H_20)	3600 μL[e]			
Mouse #	**BW (g)**	**Dilution Factor**[f]	**Aldo (μL)**[g]	**H2O (μL)**[h]
#1	30.5	0.805	241	59
#2	31.8	0.839	252	48
#3	31.2	0.823	247	53
#4	34.2	0.902	271	29
#5	33.5	0.884	265	35
#6	35.9	0.947	284	16
#7	36.8	0.971	291	9
#8	37.2	0.982	294	6
#9	37.9	1.000	300	0
#10	33.3	0.879	264	36
#11	34.1	0.900	270	30
#12	33.2	0.876	263	37

[a] $139 \times 10^{-3} \times (37.9 + ½ \times 1) \times 60 = 320.56$ ng
[b] 320.56/0.25 = 1281.024 ng
[c] 1281.024 ng × 300 = 0.384 mg/300 μL
[d] 0.384 mg × 12 = 4.608 mg
[e] 300 × 12 = 3600 μL
[f] BW/37.9 = Dilution factor
[g] 300 × dilution factor = Aldo (μL)
[h] 300 – Aldo = H_2O (μL)

numbered plastic tube. It is important to avoid drawing air into the syringe.

3. Gently insert the filling needle into the pump body. Push the syringe plunger slowly to fill the pump with Aldo solution. Stop filling the pump and carefully remove the needle as soon as a bead of fluid rises out of the pump.
4. Insert flow moderator into pump through the hole and weigh the filled pump.
5. Calculate Filling Ratio (%) = (Pump Weight "filled" – "empty") × 1000/mean fill volume × 100.
6. Place filled pump into the labeled 15 mL tube with the moderator head facing upward. Add 5 mL of sterile saline or PBS to cover the pump.
7. Place tubes in a 37 °C incubator overnight to allow the pump to prime.

3.5 Preparation for Pump Implantation and Surgical Procedures

1. Mice are anesthetized by intraperitoneal injection of ketamine (100 mg/kg) and xylazine (10 mg/kg; *see* **Note 5**).
2. An about 1 cm incision is made just behind the neck of animals (mid-scapular incision). The incision is perpendicular to the long axis of the implant (*see* **Note 9**).
3. A hemostat is inserted into the incision to create a subcutaneous pocket for the pump. Open the jaws to bluntly dissect the subcutaneous tissue. Withdraw the hemostat slightly, close jaws, and repeat the blunt dissection steps until a suitable implant pocket has been created (*see* **Note 10**).
4. The pump is inserted into the pocket, delivery portal first, which minimizes interaction between the compound delivered and the healing of the incision (*see* **Note 11**).
5. The wound is closed with wound clips (two clips will normally suffice).
6. Perform blood pressure measurement, ultrasound imaging, and quantification of aortic aneurysms as described in Subheading 3.1.

4 Notes

1. C57BL/6 mice have at least two different strains: C57BL/6J (the Jackson Laboratory) and C57BL/6N (the NIA aged rodent colonies), which have different genetic backgrounds [15]. C57BL/6J and C57BL/6N mice may potentially respond to DOCA or Aldo plus high salt with different sensitivity, which we have not extensively investigated. However, we found that there is no difference regarding induction of

aortic aneurysm by DOCA or Aldo plus high salt. In addition, the NIA aged rodent colonies are only available for projects that are directly related to aging and that are funded by grants from the NIH, other US government agencies, and US-based private foundations that fund aging research. Contract-funded projects are only eligible if they are funded by the NIA.

2. The age of C57BL/6J retired breeders from the Jackson Laboratory may be variable. Since DOCA or Aldo plus high salt model described here is age dependent [11], the retired breeders that are less than 10 months old are required for singly cage housing due to their aggression until 10 months old. Alternatively, 6-month-old C57BL/6J mice are available from the Jackson Laboratory. But again these mice are also required for singly cage housing up to 10 months of age prior to experiments.
3. Different dose (0.001–200 mg/pellet) and time release (21–90 days) of DOCA pellets are available. We use DOCA pellets at a dose of 50 mg/pellet for a 21-day release and demonstrated that it is sufficient to potently induce hypertension and aortic aneurysm in 10-month-old male C57BL/6J mice. However, it should be pointed out that the low dose of DOCA pellets may be also effective.
4. There are many different way to solubilize Aldo to use Alzet® osmotic pumps to deliver Aldo to mice (http://www.alzet.com/research_applications/ALDO.html). We used 50% DMSO and demonstrated that 50% DMSO alone did not have any effect on aortic aneurysm formation [11].
5. Mice may need to be redosed with 25–35% of the original dose of ketamine alone to achieve desirable anesthetic effect. Alternatively, mice could be anesthetized by inhalation of isoflurane mixed with O_2 (3–5% isoflurane/97% O_2) at a flow rate of 600 mL/min and anesthesia is then maintained by inhalation of isoflurane mixed with O_2 (1–3% isoflurane/97% O_2) at a flow rate of 600 mL/min throughout the surgical procedure. During anesthesia, the animal's eyes are protected with petroleum ophthalmic ointment.
6. Upon administration of DOCA and high salt, mice usually drink considerable amounts of water and urinate frequently, which indicates the effectiveness of DOCA and high salt. Personnel need to monitor and refill drinking water if necessary and change cage bedding at least every other day.
7. Increases in blood pressure by DOCA and high salt compared to basal blood pressure are as expected and indicate the effectiveness of DOCA and high salt.

8. The protocol described here uses the example of infusion of Aldo (200 μg/kg/day or 139 ng/kg/min) for 4 weeks to twelve 10-month-old male mice (Table 1).
9. Implants are best tolerated in the dorsal region of the body slightly caudal (posterior) to the shoulders. An appropriately inserted implant in this site will not interfere with vital organs, or restrict normal activities of the animal. This site is also relatively inaccessible to manipulation by the animal. Care must be exercised to cut only the skin, and not the underlying tissues.
10. The pocket will be slightly wider than the implant diameter and about 1 cm longer than the pump. Avoid making the pocket too large, as this will allow the pump to turn around or slip down on the flank of the animal.
11. The pump location is checked to assure that it does not rest immediately beneath the incision, which could interfere with the healing of the incision.

Acknowledgment

This work was supported by NIH grants HL106843 and HL125228 (to M.C. Gong and Z. Guo) and VA Merit Award BX002141 (to Z. Guo) as well as a National Institute of General. Medical Sciences grant (P20 GM103527-05 to L. Cassis).

References

1. Hiratzka LF, Bakris GL, Beckman JA et al (2010) 2010 ACCF/AHA/AATS/ACR/ASA/SCA/SCAI/SIR/STS/SVM guidelines for the diagnosis and management of patients with thoracic aortic disease. A report of the American College of Cardiology Foundation/American Heart Association task force on practice guidelines, American Association for Thoracic Surgery, American College of Radiology, American Stroke Association, Society of Cardiovascular Anesthesiologists, Society for Cardiovascular Angiography and Interventions, Society of Interventional Radiology, Society of Thoracic Surgeons, and Society for Vascular Medicine. J Am Coll Cardiol 55:e27–e129
2. Isselbacher EM (2005) Thoracic and abdominal aortic aneurysms. Circulation 111:816–828
3. Lindsay ME, Dietz HC (2011) Lessons on the pathogenesis of aneurysm from heritable conditions. Nature 473:308–316
4. Golledge J, Muller J, Daugherty A et al (2006) Abdominal aortic aneurysm: pathogenesis and implications for management. Arterioscler Thromb Vasc Biol 26:2605–2613
5. Cassis LA, Gupte M, Thayer S et al (2009) ANG II infusion promotes abdominal aortic aneurysms independent of increased blood pressure in hypercholesterolemic mice. Am J Physiol Heart Circ Physiol 296:H1660–H1665
6. Cassis LA, Helton MJ, Howatt DA et al (2005) Aldosterone does not mediate angiotensin II-induced atherosclerosis and abdominal aortic aneurysms. Br J Pharmacol 144:443–448
7. Daugherty A, Manning MW, Cassis LA (2000) Angiotensin II promotes atherosclerotic lesions and aneurysms in apolipoprotein E-deficient mice. J Clin Invest 105:1605–1612
8. Rateri DL, Moorleghen JJ, Balakrishnan A et al (2011) Endothelial cell-specific deficiency of Ang II type 1a receptors attenuates Ang II-induced ascending aortic aneurysms in LDL receptor−/− mice. Circ Res 108:574–581
9. Thomas M, Gavrila D, Mccormick ML et al (2006) Deletion of p47phox attenuates angiotensin II-induced abdominal aortic aneurysm formation in apolipoprotein E-deficient mice. Circulation 114:404–413

10. Zhang X, Thatcher SE, Rateri DL et al (2012) Transient exposure of neonatal female mice to testosterone abrogates the sexual dimorphism of abdominal aortic aneurysms. Circ Res 110:e73–e85
11. Liu S, Xie Z, Daugherty A et al (2013) Mineralocorticoid receptor agonists induce mouse aortic aneurysm formation and rupture in the presence of high salt. Arterioscler Thromb Vasc Biol 33:1568–1579
12. Weber KT (2001) Aldosterone in congestive heart failure. N Engl J Med 345:1689–1697
13. He BJ, Joiner ML, Singh MV et al (2011) Oxidation of CaMKII determines the cardiotoxic effects of aldosterone. Nat Med 17:1610–1618
14. Barisione C, Charnigo R, Howatt DA et al (2006) Rapid dilation of the abdominal aorta during infusion of angiotensin II detected by noninvasive high-frequency ultrasonography. J Vasc Surg 44:372–376
15. Mekada K, Abe K, Murakami A et al (2009) Genetic differences among C57BL/6 substrains. Exp Anim 58:141–149

Chapter 13

Fluorescence-Based Binding Assay for Screening Ligands of Angiotensin Receptors

Maiia E. Bragina, Nikolaos Stergiopulos, and Rodrigo A. Fraga-Silva

Abstract

Binding assay is a common technique used to characterize ability of a ligand to interact with a specific biological target. A number of parameters, such as binding affinity, receptor density, and association/dissociation rate constants, can be measured by means of this technique. In most cases, implementation of the binding assay requires specific infrastructure for labeling and detecting the ligand, which impedes realization of this technique in a standard laboratory. Here we describe a simple fluorescence-based binding assay for angiotensin peptides and receptors, which does not require complex equipment and can be used for initial screening of the novel ligands or mutational studies.

Key words Binding assay, Ligand, Receptor, Fluorescently labeled, Screening, Renin–angiotensin system

1 Introduction

The relevance of the renin angiotensin system (RAS) is demonstrated by the remarkable clinical success of the angiotensin type I receptor (AT1) blockers and angiotensin converting enzyme (ACE) inhibitors [1–3]. Nowadays, drugs targeting RAS represent one of the most effective class of medications used for treatment of cardiovascular diseases [1, 2]. Recent discovery of new RAS components and their essential role in various pathophysiological processes pointed out novel pharmacological targets. For instance, newly identified G-protein-coupled receptors (GPCRs), AT2 and Mas, activated by angiotensin II and angiotensin-(1–7), respectively, have been shown to evoke a number of effects opposite to ones associated with AT1 signaling [4–6]. Thus compounds, which are able to activate these receptors may become potential therapeutic tools against cardiovascular disorders [5].

Binding of ligand to the receptor is a fundamental step, responsible for modulation of the receptor activity and downstream signaling, which in turn results in development of various physiological,

Sean E. Thatcher (ed.), *The Renin-Angiotensin-Aldosterone System: Methods and Protocols*, Methods in Molecular Biology, vol. 1614, DOI 10.1007/978-1-4939-7030-8_13, © Springer Science+Business Media LLC 2017

pathological, or pharmacological effects [7, 8]. Binding assay is a prominent technique used to detect and characterize ligand–receptor interactions for elucidation of molecular regulatory mechanisms [4, 9] as well as development of novel drugs [10, 11] or diagnostic tools [12, 13]. Although a wide variety of methods for realization of binding assays have been established, in most cases this assay consists of incubation of labeled ligand with the corresponding receptor preparation and following quantification of the ligand bound to the receptor [8].

Specific parameter to be determined by means of binding assay defines the experimental design. In this sense three types of binding assay can be highlighted: saturation binding assay, kinetic studies, and competition binding. In particular, saturation experiments are used to assess receptor density (B_{max}) and affinity of the ligand to the receptor (K_d) [14, 15]. Generally, it involves incubation of receptor preparation with increasing concentrations of labeled ligand [15]. Ligand bound to the receptor is further quantified and plotted against its concentration in the assay mixture [14, 15]. On its turn, kinetic studies are generally employed to measure the association/dissociation (k_{on}/k_{off}) rate constants of the binding reaction and dissociation half-life ($t_{½}$) [16]. In order to evaluate the association kinetics, receptor preparation is incubated with labeled ligand at one or more fixed concentrations. At the same time, dissociation studies are performed by pre-incubation of receptors with labeled ligand until equilibrium is reached and subsequent initiation of the dissociation process by addition of unlabeled competitor or immersion in the buffer solution [15]. Both reactions are allowed to run for different time periods and consequently intensity of the signal, created by the label, is plotted with respect to the time point [15].

Among all types of binding assay, competition binding is the most common one used to characterize binding features of angiotensin peptides. Competition experiment tests the ability of unlabeled ligand to compete for the binding site of the receptor with the labeled ligand of fixed concentration [15]. Competing ligand can be identical to the labeled compound (labeled vs. non-labeled) or be structurally different (e.g., synthetic agonist/antagonist vs. endogenous ligand) [14]. In case of the identical ligands used, competitive binding assay can be employed as an alternative to saturation studies to assess K_d and B_{max} [14]. Experiment with structurally different ligands gives rise to the concentration of the unlabeled compound, which displaces 50% of the labeled ligand bound to the target receptor (IC_{50}) [8]. Final result of the competition binding assay is the curve of displacement, where binding of labeled ligand is plotted with respect to log concentration of the non-labeled ligand of interest [14]. Binding of labeled compound is expressed as a fraction of total binding, which arises from incubation of the receptor preparation with the labeled ligand only [14].

Ligand labeling strategy is the most important parameter of the binding assay as it impacts the whole experimental procedure. Due to its sensitivity and specificity, radioligand binding assay, which implies incorporation of the radioactive moiety into the ligand molecule, has become a conventional technique used to characterize ligand–receptor interactions. However, employment of this method is associated with a number of drawbacks like necessity to handle radioactive substances as well as production of hazardous waste [8].

In the current chapter we describe a simple binding assay technique, which is based on usage of fluorescently labeled ligands. While being less accurate as the radioligand methods, it is safer, has lower costs and requires standard laboratory equipment. Although not accurate enough to determine K_d or B_{max}, the technique introduced in this chapter can serve as a valuable tool to rapidly test large number of compounds for their ability to bind the receptor. Another useful application of the current method is mutational studies to identify specific amino acids within the binding site of the receptor.

2 Materials

1. Cells naturally expressing the target receptor or transfected cells. A number of approaches are currently available for delivery of foreign genetic material into the cell: biological (virus-mediated), chemical (cationic polymers and lipids, calcium phosphate), and physical (direct injection, electroporation, etc.) [17]. One should select the method, which will ensure high transfection efficiency, reproducibility, and protein yield (*see* **Note 1**).
2. Assay solution: Hanks' Balanced Salt solution or standard Tris buffer supplemented with 0.2% bovine serum albumin (BSA), 0.01% bacitracin, 0.002% phenylmethylsulfonyl fluoride (PMSF), 0.01% 1,10-phenanthroline (*see* **Notes 2** and **3**). To reduce background fluorescence Hanks' solution without phenol red is preferred. Since 1,10-phenanthroline and PMSF have only moderate solubility in water, it is recommended to pre-solubilize them in methanol and isopropanol, respectively. For 100 mL of assay solution, dilute 2 mg of PSMF in 150 μL of isopropanol and 10 mg of 1,10-phenanthroline in 50 μL of methanol. Final formulation of the assay solution must be done just before the experiment, since the supplementary components, PMSF in particular, have limited stability in aqueous solutions.
3. Angiotensin peptides or testing compounds in unlabeled form as a competing ligand, for example: (a) Angiotensin II-Asp-Arg-Val-Tyr-Ile-His-Pro-Phe; (b) Angiotensin-(1–7)-Asp-Arg-Val-Tyr-Ile-His-Pro; (c) Angiotensin-A-Ala-Arg-Val-Tyr-Ile-His-Pro-Phe;

(d) Alamandine-Ala-Arg-Val-Tyr-Ile-His-Pro; (e) Angiotensin-(1–9)-Asp-Arg-Val-Tyr-Ile-His-Pro-Phe-His; and (f) Angiotensin IV-Val-Tyr-Ile-His-Pro-Phe. In case binding properties of the competing ligand are not known, we recommend to test several concentrations of it in order to cover the condition of its possible interaction with the target receptor. Assay solution containing different amount of the competing ligand should be prepared right before the experiment and kept on ice.

4. Fluorescently labeled angiotensin peptide (labeled with Mono-5-(and-6)-carboxyfluorescein (FAM) or Mono-Rhodamine B (Rhod), e.g.: FAM-Angiotensin II; Rhod-Angiotensin II; FAM-Angiotensin-(1–7); Rhod-Angiotensin-(1–7)). Dilute labeled peptide in the assay solution right before the experiment initiation and keep it protected from light and on ice (*see* **Note 4**).
5. Receptor antagonists can also be used in certain experiment design (e.g.: losartan, AT1 antagonist; A-779, Mas antagonist; PD123319, AT2 antagonist) (*see* **Note 5**). Similar to the peptide preparation, dilute the antagonist in the assay buffer, right before the experiment initiation and keep it on ice.
6. Twelve-well plate with coverslips fitting into the well (e.g., 15 mm circular coverslip) or any other support/flask, which allows for cell cultivation and consequent microscopy of the samples.
7. Glass slides (e.g., 25 × 75 × 1.0 mm), DAPI containing mounting medium (*see* **Note 6**) and nail polish to prepare coverslips with cells for microscopy.
8. Fine tweezers to manipulate the coverslips and mount them on the glass slides.
9. Fluorescent or confocal microscope for image acquisition.
10. Ice to keep assay components and cell preparation at low temperature.

Diligently follow all waste disposal regulations when discarding waste materials.

3 Method

The current chapter describes a method to test the ability of the compound of interest to bind to angiotensin-related GPCRs [18–20].

Cells constitutively expressing the GPCR of interest (target receptor) should be seeded on coverslips, placed in the 12-well plate, 1–2 days in advance of the binding assay (*see* **Note 7**). In case transiently transfected cells are used in the experiment, duration of the cell cultivation should correspond to the maximum level of the receptor expression (*see* **Note 1**).

When conducting binding experiments one should distinguish between specific and nonspecific binding of the ligand [7]. While specific binding (SB) implies association of the ligand exclusively with the target receptor, nonspecific binding (NSB) results from the labeled ligand binding to the other components present in the assay mixture. Since NSB may introduce a significant disturbance in the binding measurements, it is important to quantify it in every binding experiment. NSB is determined in the presence of excess of the unlabeled ligand or receptor antagonist, which leads to the occupation of the target receptors by non-detectable molecule and labeled ligand binding mostly to the nonspecific sites. Although it is advisable to have NSB no more than 20% [14], significant results can be still obtained at NSB level up to 30–60% [9, 20, 21].

All the procedures should be performed on ice to minimize the protease activity and inhibit a range of the processes such as internalization of the receptor [7].

We recommend to run triplicates for each assay condition.

1. Once the monolayer of cells expressing the target receptor is formed on the coverslips, aspirate the medium and wash twice with 1–2 mL of the ice-cold assay solution. This step should be done gently to avoid damage of the cell monolayer.
2. Incubate cells for 5–10 min with prepared dilutions of the testing compound or unlabeled angiotensin (at designated concentrations) to occupy the binding sites of the target receptor. For total binding measurements, the pre-incubation must be performed using assay buffer without testing compound or unlabeled angiotensin peptide, while nonspecific binding measurements should include pre-incubation with high concentration of unlabeled angiotensin peptide (e.g., 10^{-5} M). We recommend to add 1 mL of the solution per well in case of 12-well plate (*see* **Note 8**).
3. Gently wash the cells with the assay solution.
4. Add the solution containing unlabeled peptide or testing compound at same concentrations as for pre-incubation step plus labeled angiotensin peptide into the wells. The concentration of labeled ligand must be the same in every experimental condition. Let the incubation last for 60 min on ice with gentle shaking every 10–15 min. From this step on, the preparation is photosensitive and must be protected from light exposure.
5. After incubation, wash gently 2–3 times with ice-cold assay solution (*see* **Note 9**).
6. Place a small drop of the mounting medium on the microscopy slide. Using the fine tweezers, take the coverslip from the well, carefully remove the excess of liquid without touching the cell

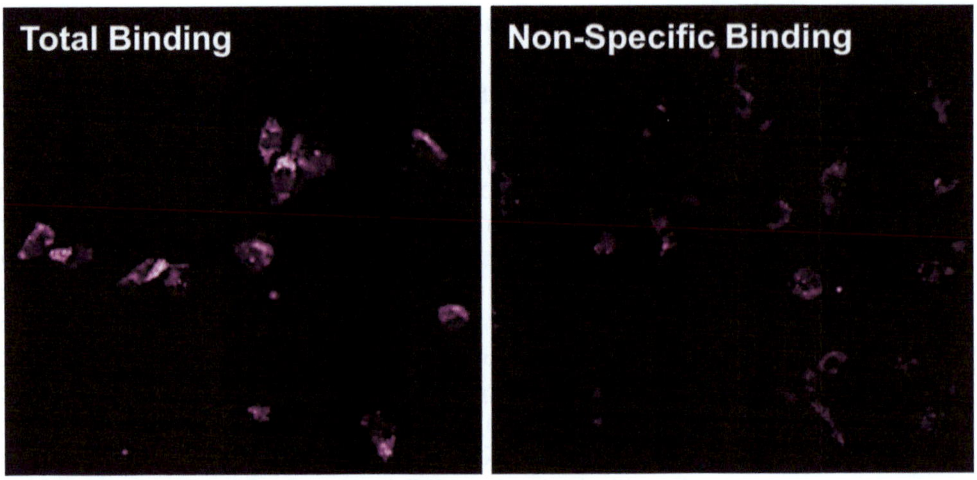

Fig. 1 Representative images corresponding to conditions of (**a**) total binding and (**b**) nonspecific binding of FAM-angiotensin-(1–7) to CHO cells overexpressing Mas receptor

monolayer and subsequently mount the coverslip with cells down on the microscope slide. Seal the edges of coverslip using, for example, nail polish. Do not forget to properly label the microscope slide.

Of note, it could be problematic to remove the coverslip from the well in wet environment. The selection of a suitable tweezer with pointed tips can facilitate the coverslip handling.

7. Directly proceed with imaging by means of fluorescent or confocal microscope (*see* **Note 10**). As images are taken for the following signal quantification, all acquiring parameters must be kept unchanged during imaging. It is recommended to capture several pictures (at least 5) from every slide for better overview of each condition (Fig. 1).
8. Quantification can be done with different image processing software, such as Image J (NIH, Bethesda, MD, USA). The main principle of the image analysis is to quantify the mean grey value (signal intensity) on every image and normalize it to the number of cells present in the field of view. The number of cells can be obtained by counting number of nuclei stained by DAPI. Artefacts exhibiting amorphous strong fluorescent signal should be excluded from the area of quantification.

4 Notes

1. Since the introduced fluorescence-based method has low sensitivity, we highly recommend to use cells, which overexpress the receptor of interest. In case transfection is done to achieve the receptor overexpression, it is advisable to carry out preliminary

optimization of the transfection conditions as well as duration of cell cultivation to achieve the highest possible receptor density. Although radioligand-based methods allow usage of tissue sections for binding studies, we do not recommend to employ them with the current technique. Tissues often present high level of background fluorescence, which can produce considerable noise, impairing the sensitivity of the experiment. Exceptions can be made in case tissue has high expression of the target receptor, as shown in previous publications assessing binding of Angiotensin-(1–7) to Mas receptor in mouse testis [22] and heart [23] sections.

2. Angiotensin peptides have a very short half-life and enzymes, constitutively expressed by the cultured cells, can cause their rapid degradation during the experiment. In order to increase the stability of the peptides, protease inhibitors (bacitracin, PMSF, and 1,10-phenanthroline) must be added to the assay solution. Additionally, BSA should be added in the assay solution to prevent nonspecific adsorption of the peptide onto surfaces.

3. pH of the solution is a critical parameter, which can significantly change the ligand–receptor binding affinities through disturbance of electrostatic/hydrophobic interactions as well as receptor conformation. The pH of the assay mixture must correspond to the physiological range and stay constant during the experiment, and therefore, buffer should be added into the composition of assay solution. We recommend to use standard Tris buffer (e.g., 150 mM NaCl, 50 mM Tris–HCl, 20 mM EDTA) with pH adjusted to 7.4, or Hanks' solution, which does not contain sodium bicarbonate, but instead buffered by 20 mM HEPES. While the bicarbonate-buffered solutions tend to alkalinize at low CO_2 content in the ambient atmosphere, HEPES keeps the pH constant. In some cases, calcium and/or magnesium salts could be added to promote the binding events [14].

4. Competition binding assays require labeled ligand concentrations equal or less than K_d for the given receptor–ligand pair [14] since higher concentrations may be not displaced by the competitive ligand. As an example, for FAM-labeled angiotensin-(1–7) the optimal concentration is 10^{-8}–10^{-9} M [20, 21]. In case of limited information about binding properties of the labeled ligand, additional optimization of its concentration may be required.

5. Antagonist of high affinity and selectivity (e.g., losartan, AT1 antagonist; A-779, Mas antagonist; PD123319, AT2 antagonist) may be employed at high concentration to additionally confirm nonspecific binding [21].

6. We recommend using DAPI-containing mounting medium to facilitate imaging of the samples and cell quantification during the image analysis.
7. It is advisable that cells have the confluence of no more than 60–70% by the day of the experiment since monolayer of higher density may complicate cell quantification and overall image analysis.
8. In order to use the current technique for the mutational studies, conditions with (a) fluorescently labeled angiotensin (total binding) and (b) excess of unlabeled form co-incubated with fluorescently labeled angiotensin (nonspecific binding) are sufficient to drive the conclusion whether the angiotensin peptide is able to bind to the mutated receptor. Difference between total binding and nonspecific binding will indicate the binding event.
9. The final washing step is critical to guarantee the removal of the unbound labeled ligands, which can significantly increase the background fluorescence. However, taking into account that removal of the ligand-containing solution disturbs the equilibrium and triggers the dissociation of the ligand, we advise that final washing step and subsequent preparation of the microscope slide are done as fast as possible. Please refer to Fig. 2 for processing microscope slides.
10. For image acquisition, it is recommended to use high resolution fluorescent microscope, equipped with appropriated filters, or confocal microscope. However, when using confocal, an adequate opening of the pinhole must be considered.

Conditions \ Steps	1 *Washing*	2 *Pre-Incubation*	3 *Washing*	4 *Incubation*	5 *Washing*	6 *Mounting*	7 *Imaging*	8 *Analysis*
Total Binding	**Wash the cells with assay solution**	Assay Solution *(without ligand)*	**Wash the cells with assay solution**	FLP	**Wash the cells with assay solution**	Mount the coverslip on the slide	Image acquisition (fluorescent or confocal microscopy)	Image analysis (intensity / cell number)
Non-Specific Binding		*UAP*		*FLP + UAP*				
Binding of Testing Compound *(at different concentration)*		TC		FLP + TC				

UAP - unlabeled angiotensin peptide (diluted in assay buffer)
FLP - fluorescently labeled peptide (diluted in assay buffer)
TC - testing compound (diluted in assay buffer)

Fig. 2 Schematic process-flow

References

1. Schiffrin EL (2002) Vascular and cardiac benefits of angiotensin receptor blockers. Am J Med 113(5):409–418
2. Ma TK, Kam KK, Yan BP, Lam YY (2010) Renin-angiotensin-aldosterone system blockade for cardiovascular diseases: current status. Br J Pharmacol 160(6):1273–1292. doi:10.1111/j.1476-5381.2010.00750.x
3. Bader M (2010) Tissue renin-angiotensin-aldosterone systems: targets for pharmacological therapy. Annu Rev Pharmacol Toxicol 50:439–465. doi:10.1146/annurev.pharmtox.010909.105610
4. Santos RA, Simoes e Silva AC, Maric C, Silva DM, Machado RP, de Buhr I, Heringer-Walther S, Pinheiro SV, Lopes MT, Bader M, Mendes EP, Lemos VS, Campagnole-Santos MJ, Schultheiss HP, Speth R, Walther T (2003) Angiotensin-(1–7) is an endogenous ligand for the G protein-coupled receptor mas. Proc Natl Acad Sci U S A 100(14):8258–8263. doi:10.1073/pnas.1432869100
5. Fraga-Silva RA, Ferreira AJ, Dos Santos RA (2013) Opportunities for targeting the angiotensin-converting enzyme 2/angiotensin-(1–7)/mas receptor pathway in hypertension. Curr Hypertens Rep 15(1):31–38. doi:10.1007/s11906-012-0324-1
6. Steckelings UM, Paulis L, Namsolleck P, Unger T (2012) AT2 receptor agonists: hypertension and beyond. Curr Opin Nephrol Hypertens 21(2):142–146. doi:10.1097/MNH.0b013e328350261b
7. Sridharan R, Zuber J, Connelly SM, Mathew E, Dumont ME (2014) Fluorescent approaches for understanding interactions of ligands with G protein coupled receptors. Biochim Biophys Acta 1838(1 Pt A):15–33. doi:10.1016/j.bbamem.2013.09.005
8. de Jong LA, Uges DR, Franke JP, Bischoff R (2005) Receptor-ligand binding assays: technologies and applications. J Chromatogr B Anal Technol Biomed Life Sci 829(1–2):1–25. doi:10.1016/j.jchromb.2005.10.002
9. Lautner RQ, Villela DC, Fraga-Silva RA, Silva N, Verano-Braga T, Costa-Fraga F, Jankowski J, Jankowski V, Sousa F, Alzamora A, Soares E, Barbosa C, Kjeldsen F, Oliveira A, Braga J, Savergnini S, Maia G, Peluso AB, Passos-Silva D, Ferreira A, Alves F, Martins A, Raizada M, Paula R, Motta-Santos D, Klempin F, Pimenta A, Alenina N, Sinisterra R, Bader M, Campagnole-Santos MJ, Santos RA (2013) Discovery and characterization of alamandine: a novel component of the renin-angiotensin system. Circ Res 112(8):1104–1111. doi:10.1161/CIRCRESAHA.113.301077
10. Violin JD, DeWire SM, Yamashita D, Rominger DH, Nguyen L, Schiller K, Whalen EJ, Gowen M, Lark MW (2010) Selectively engaging beta-arrestins at the angiotensin II type 1 receptor reduces blood pressure and increases cardiac performance. J Pharmacol Exp Ther 335(3):572–579. doi:10.1124/jpet.110.173005
11. Wan Y, Wallinder C, Plouffe B, Beaudry H, Mahalingam AK, Wu X, Johansson B, Holm M, Botoros M, Karlen A, Pettersson A, Nyberg F, Fandriks L, Gallo-Payet N, Hallberg A, Alterman M (2004) Design, synthesis, and biological evaluation of the first selective nonpeptide AT2 receptor agonist. J Med Chem 47(24):5995–6008. doi:10.1021/jm049715t
12. Slavik R, Herde AM, Bieri D, Weber M, Schibli R, Kramer SD, Ametamey SM, Mu L (2015) Synthesis, radiolabeling and evaluation of novel 4-oxo-quinoline derivatives as PET tracers for imaging cannabinoid type 2 receptor. Eur J Med Chem 92:554–564. doi:10.1016/j.ejmech.2015.01.028
13. Ni R, Gillberg PG, Bergfors A, Marutle A, Nordberg A (2013) Amyloid tracers detect multiple binding sites in Alzheimer's disease brain tissue. Brain 136(Pt 7):2217–2227. doi:10.1093/brain/awt142
14. Auld DS, Farmen MW, Kahl SD, Kriauciunas A, McKnight KL, Montrose C, Weidner JR (2012) Receptor binding assays for HTS and drug discovery. In: Sittampalam GS, Coussens NP et al (eds) Assay guidance manual. Eli Lilly & Company and the National Center for Advancing Translational Sciences, Bethesda, MD
15. Maguire JJ, Kuc RE, Davenport AP (2012) Radioligand binding assays and their analysis. In: Davenport AP (ed) Receptor binding techniques, vol 897, 3rd edn. Springer, New York, NY
16. Hulme EC, Trevethick MA (2010) Ligand binding assays at equilibrium: validation and interpretation. Br J Pharmacol 161(6):1219–1237.doi:10.1111/j.1476-5381.2009.00604.x
17. Kim TK, Eberwine JH (2010) Mammalian cell transfection: the present and the future. Anal Bioanal Chem 397(8):3173–3178. doi:10.1007/s00216-010-3821-6
18. Jankowski V, Tolle M, Santos RA, Gunthner T, Krause E, Beyermann M, Welker P, Bader M, Pinheiro SV, Sampaio WO, Lautner R, Kretschmer A, van der Giet M, Zidek W, Jankowski J (2011) Angioprotectin: an angiotensin II-like peptide causing vasodilatory effects. FASEB J 25(9):2987–2995. doi:10.1096/fj.11-185470
19. Pinheiro SV, Simoes e Silva AC, Sampaio WO, de Paula RD, Mendes EP, Bontempo ED, Pesquero

JB, Walther T, Alenina N, Bader M, Bleich M, Santos RA (2004) Nonpeptide AVE 0991 is an angiotensin-(1–7) receptor mas agonist in the mouse kidney. Hypertension 44(4):490–496. doi:10.1161/01.HYP.0000141438.64887.42

20. Savergnini SQ, Beiman M, Lautner RQ, de Paula-Carvalho V, Allahdadi K, Pessoa DC, Costa-Fraga FP, Fraga-Silva RA, Cojocaru G, Cohen Y, Bader M, de Almeida AP, Rotman G, Santos RA (2010) Vascular relaxation, antihypertensive effect, and cardioprotection of a novel peptide agonist of the MAS receptor. Hypertension 56(1):112–120. doi:10.1161/HYPERTENSIONAHA.110.152942
21. Fraga-Silva RA, Pinheiro SV, Goncalves AC, Alenina N, Bader M, Santos RA (2008) The antithrombotic effect of angiotensin-(1–7) involves mas-mediated NO release from platelets. Mol Med 14(1–2):28–35. doi:10.2119/2007-00073.Fraga-Silva
22. Leal MC, Pinheiro SV, Ferreira AJ, Santos RA, Bordoni LS, Alenina N, Bader M, Franca LR (2009) The role of angiotensin-(1–7) receptor mas in spermatogenesis in mice and rats. J Anat 214(5):736–743. doi:10.1111/j.1469-7580.2009.01058.x
23. Santos RA, Castro CH, Gava E, Pinheiro SV, Almeida AP, Paula RD, Cruz JS, Ramos AS, Rosa KT, Irigoyen MC, Bader M, Alenina N, Kitten GT, Ferreira AJ (2006) Impairment of in vitro and in vivo heart function in angiotensin-(1–7) receptor MAS knockout mice. Hypertension 47(5):996–1002. doi:10.1161/01.HYP.0000215289.51180.5c

Chapter 14

A Primer to Angiotensin Peptide Isolation, Stability, and Analysis by Nano-Liquid Chromatography with Mass Detection

Mariola Olkowicz, Stefan Chlopicki, and Ryszard T. Smolenski

Abstract

The renin-angiotensin system (RAS) is an important element of cardiovascular and renal physiology and targeting the RAS by renin inhibitors, angiotensin (Ang) converting enzyme (ACE) inhibitors and Ang II type 1 receptor antagonists is effective in the treatment of hypertension, heart failure, and atherosclerosis. Quantification of Ang peptides is critical to establish the status of the RAS, but it is challenging due to low Ang peptides concentrations (fmol/mL or fmol/g), abundance of interfering substances, post sampling conversions, and difficulties with the specificity of the assay.

In this chapter, we describe a new nano-LC/MS-based methodology for comprehensive, specific, sensitive, and accurate quantification of Ang peptides profile in plasma and tissue. We optimized sample pretreatment method (protein removal (acetonitrile precipitation) followed by solid-phase extraction (C18 silica bonded phase)), chromatographic conditions (reversed-phase nanochromatography with preconcentration), and mass detection (multiple reaction monitoring) of nine peptides: Ang-(1–12), Ang I (1–10), Ang-(1–9), Ang II (1–8), [Ala^1]-Ang II, Ang III (2–8), Ang IV (3–8), Ang-(1–7), and [Ala^1]-Ang-(1–7). Assessment of plasma and cardiac concentrations of Ang peptides in genetically modified atherosclerotic apolipoprotein E/LDL receptor double knockout (ApoE$^{-/-}$/LDLR$^{-/-}$) mice vs. wild types revealed changes in renin-angiotensin system consistent with an overactivation of ACE and impairment of ACE2. The method could be easily adopted for high-throughput analysis and for use in clinical applications such as diagnosis of the RAS abnormalities or monitoring of the RAS inhibition-based therapies.

Key words Renin-angiotensin system, Angiotensin peptides, Sample handling and extraction, Nanoflow liquid chromatography, Multiple reaction monitoring, Angiotensin concentrations in plasma and tissues

1 Introduction

The renin-angiotensin system (RAS) has been originally characterized as a set of circulating factors that play a critical role in cardiovascular and renal physiology [1, 2]. Further research revealed that disturbances of this system lead to hypertension, atherosclerosis, cardiac hypertrophy, heart failure, and renovascular disorders [3–5].

Sean E. Thatcher (ed.), *The Renin-Angiotensin-Aldosterone System: Methods and Protocols*, Methods in Molecular Biology, vol. 1614, DOI 10.1007/978-1-4939-7030-8_14, © Springer Science+Business Media LLC 2017

Angiotensin II (Ang II), the primary effector of the RAS is involved not only in control of blood pressure but is also affecting the function of most of the organs (including heart, blood vessels, kidney, and brain) causing mostly deleterious effects. On the other hand, the counterregulatory axis of the RAS, centered on the actions of angiotensin-converting enzyme 2 (ACE2), and the resultant Ang-(1–7) and Ang-(1-9) might oppose some harmful effects of Ang II [6–9]. Moreover, newly identified RAS family peptides: Ang A and alamandine further extend the complexities in understanding the cardiovascular/renal physiopathology of the RAS [10–12]. Altogether, the RAS could be described as a complex array of components with diverse biological actions that can be functionally partitioned into distinct peptidases, peptide signaling molecules, and receptors that are downstream from the initial processing of angiotensinogen to form Ang I and Ang-(1–12) precursor peptides [13] (Fig. 1).

In view of the importance of the RAS and increased concerns on reliable quantification of endogenous angiotensin peptides in various biological compartments, a new approach is needed to accurately characterize the status of this system. The immunoassay-based methods such as radioimmunoassay (RIA) or enzyme-linked immunosorbent assay (ELISA) are valuable tools and currently remain the most commonly used procedure to quantitate individual angiotensins [14–17]. However, limitations of these assays

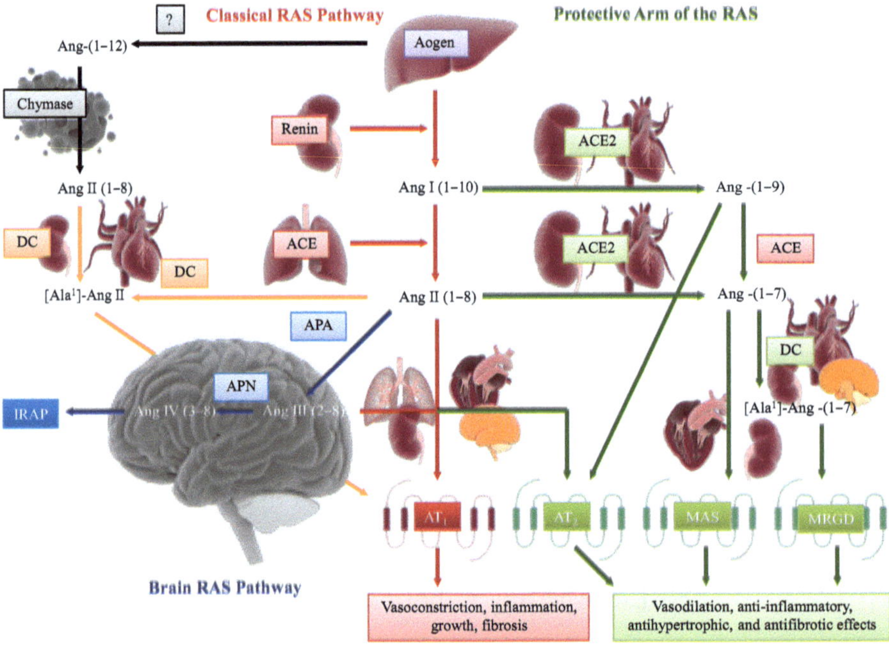

Fig. 1 Simplified view of the renin–angiotensin system (RAS) cascade. Aogen angiotensinogen, Ang angiotensin, ACE angiotensin-converting enzyme, ACE2 angiotensin-converting enzyme 2, APA aminopeptidase A, APN aminopeptidase N, DC decarboxylase, AT1 angiotensin type 1 receptor, AT2 angiotensin type 2 receptor, IRAP insulin-regulated aminopeptidase, Mas Ang-(1–7) receptor Mas, MrgD Mas related G-protein coupled receptor D

such as variable specificity that dependent on the characteristics for each antibody, laborious and time-consuming procedures as well as problems with recovery assessment made search for new technologies a high priority. An alternative approach is based on application of liquid chromatography/mass spectrometry (LC/MS) that enables detecting peptide fragmentation signatures/mass transitions that are highly specific for individual angiotensins. This in combination with high sensitivity and high-throughput capability for sample processing and analytical step make this method ideal for such analysis [18–20].

This chapter provides protocols from the initial sample processing to angiotensin peptides analysis by nano-LC-MS/MS as well as their application in the study of the systemic and tissue RAS abnormalities. The method proposed was used for simultaneous detection and quantification of nine angiotensins in plasma and heart of wild type (C57BL/6J) and atherosclerotic ($ApoE^{-/-}$/ $LDLR^{-/-}$) mice, indicating substantial dysregulation of the RAS in these atherosclerosis-prone mice.

2 Materials

Prepare all solutions for mass spectrometry analysis using ultrapure 18 MΩ water (obtained by reverse osmosis/deionization) and analytical/MS grade reagents. MS analysis can be performed with any mass spectrometer with MS/MS capabilities. In this report concentrations and amounts of reagents and solutions were optimized for a nanoflow liquid chromatography system coupled online to a TSQ Vantage triple quadrupole mass spectrometer.

2.1 Equipment

1. Water purification system.
2. Ultrasonic bath.
3. Digital vortex mixer.
4. Eppendorf microcentrifuge.
5. SpeedVac concentrator.
6. Vac elut 20 manifold/Solid phase extraction system.
7. UltiMate 3000 Rapid Separation nano-LC system.
8. ChipMate nanoelectrospray ion source.
9. TSQ Vantage EMR triple quadrupole mass spectrometer.
10. Xcalibur version 2.1 software.

2.2 Reagents

1. Peptide standards: Ang-(1–12), Ang I (1–10), Ang-(1–9), Ang II (1–8), [Ala^1]-Ang II, Ang III (2–8), Ang IV (3–8), Ang-(1–7), [Ala^1]-Ang-(1–7) and internal standard: [Asn^1, Val^5]-Ang II (Phoenix Europe GmbH, Karlsruhe, Germany and Sigma-Aldrich, Poznan, Poland).

2. Protease Inhibitor Cocktail: 10 mM ethylenediaminetetraacetic acid (EDTA) disodium salt, 5 mM 1,10-phenanthroline, 20 mM 4-(2-aminoethyl)benzenesulfonyl fluoride (AEBSF), 400 μM amastatin, 150 μM pepstatin A, 100 μM thiorphan, and 500 μM aliskiren.
3. 0.9% solution of sodium chloride.
4. Protein Extraction Solution: 0.9% saline, 0.1 M HCl, 10% protease inhibitor cocktail (v/v).
5. BCA (bicinchoninic acid) protein assay kit.

2.3 Chromatographic Accessories and Other Materials

1. Blood collection tubes containing EDTA.
2. Low-binding 0.5/1.5-mL microcentrifuge tubes and low-retention pipette tips.
3. C18 silica bonded SPE cartridges (100 mg/1 mL), Sep-Pak type.
4. Trapping column for the nanoflow LC system: Acclaim PepMap100 C18, 2 cm × 75 μm, 3 μm, 100 Å.
5. Separation column: Acclaim PepMap100 RSLC C18, 150 mm × 75 μm I.D., 2 μm, 100 Å.

3 Methods

3.1 Determination of Plasma Angiotensin Content

The whole blood was collected via cardiac puncture into chilled tubes (containing EDTA as anticoagulant) from 8-month-old male ApoE/LDLR double knockout (n = 7) and wild type C57BL/6J mice (n = 5) under isoflurane anesthesia (*see* **Notes 1** and **2**). Afterwards, the samples were immediately mixed with protease inhibitor cocktail at a ratio of 9:1 (v/v) and centrifuged at 2000 × g, 4 °C for 20 min to isolate plasma (*see* **Notes 3–8**). After the centrifugation, the resulting plasma was transferred into the Protein LoBind tubes, split into aliquots and if not used stored at −80 °C (*see* **Note 9**).

3.1.1 Preparation of the Standard Solutions, Calibration Standards (CS), Internal Standards (IS), and Quality Control (QC) Samples

1. Prepare the stock solution of each standard at the concentration of 1 mg/mL in the amber glass vials, 50% acetonitrile (ACN)/1% acetic acid and keep frozen at −80 °C until used (*see* **Notes 10** and **11**).
2. Prepare the working standard solutions by dilution of the stock solutions with deionized water to obtain concentration in the range of 50–2500 pg/mL.
3. Mix and dilute of IS stock solution with water to obtain the IS working solution at 2000 pg/mL.
4. Prepare the calibration standards/or quality control samples just before use by spiking 70 μL (pooled blank) plasma with 20

μL appropriate working standard solution and 10 μL IS – [Asn[1], Val[5]]-Ang II (2000 pg/mL).

5. Calibration curve standards should be made in at least six levels covering the expected range. In our protocol the standards were prepared at: 5; 10; 50; 100; 250; 400; 500 pg/mL (IS – 200 pg/mL) concentrations, respectively.
6. Pretreat each of the samples in accordance with the protocol outlined below.

3.2 Sample Extraction

Sample cleanup procedure consisted of two stages: (1) protein precipitation and (2) solid-phase extraction that were performed as outlined below (*see* **Note 12**). Samples should be kept on ice unless otherwise specified.

3.2.1 Protein Precipitation

1. Mix 70 μL aliquot of each plasma sample (except for the calibration standards/or quality control samples) with 20 μL of deionized water in order to make the sample volume and condition identical with calibration standards/quality control samples.
2. Add 10 μL of the IS working solution, mix rapidly and extract with 400 μL of pure, HPLC-grade ACN.
3. Vortex the mixture and allow to stand at −20 °C for 45 min.
4. Remove the precipitated proteins by centrifugation at 20,000 × *g* for 20 min and evaporate the collected supernatant to dryness.

3.2.2 Solid-Phase Extraction

1. Dissolve the residues obtained in 50 μL 1% (v/v) acetic acid in water, vortex and sonicate two times for 2 min to facilitate peptides dissolution and then centrifuge (20,000 × *g*, 10 min, 4 °C) to separate any insoluble particles.
2. Precondition each cartridge with 2 mL ACN followed by 2 mL of water (both containing 1% (v/v) acetic acid) at a suction pressure of 5 mmHg.
3. Apply the thawed sample (~50 μL) to the cartridge and wash with 2 mL of 1% (v/v) acetic acid in water.
4. Elute the angiotensin peptides retained at the columns with 0.5 mL of H_2O/ACN, 20:80 (v/v) with 1% (v/v) acetic acid (suction pressure less than 5 mmHg) into the Protein LoBind polypropylene tubes.
5. Concentrate the eluates using a vacuum centrifuge, freeze aliquots at −80 °C and lyophilize them for 15–18 h until completely dry.
6. Reconstitute dried peptide samples with 35 μL of 1% (v/v) acetic acid in water and analyze using LC/MS system.

3.3 Analysis by Nano-LC-MS/MS

1. Perform two replicate injections (3 μL) per sample in the mobile phase – ACN/H_2O, 2:98 (v/v) containing 1% (v/v) acetic acid, at a flow rate of 5 μL/min for 5 min.
2. Load peptides onto a trapping column to preconcentrate and desalt samples.
3. Perform a reversed-phase liquid chromatography (RPLC) using UltiMate nano-LC system on a C18 column in a gradient of phase A (acetic acid (1%, v/v) in water) and phase B (acetic acid (1%, v/v) in acetonitrile).
4. Elute the peptides at a flow rate of 300 nL/min in a continuous acetonitrile gradient: 2% B for 5 min, 2–98% B for 15 min, and 98% B for 5 min.
5. Set the mass spectrometer to operate in positive ion mode and adjust other parameters depending on the type/model used. Measure two different mass transitions for each peptide including internal standard. Analytical method details as well as the MS/MS transitions monitored in our protocol are presented in Table 1 (*see* **Notes 13** and **14**).
6. Analyze the peptides quantitatively by means of the standard addition calibration.
7. In our case, Xcalibur v. 2.1 software was used for control of both, the Ultimate nano-LC and TSQ Vantage system.

Representative LC/MS chromatograms of selected angiotensin peptides and internal standard from the plasma sample of ApoE$^{-/-}$/LDLR$^{-/-}$mouse are presented on Fig. 2.

3.4 Determination of Cardiac Angiotensin Content

As in the case of determinations in plasma described above, heart specimens were obtained from 8-month-old male ApoE/LDLR double knock-out and wild type C57BL/6J mice. Hearts were rapidly removed, immediately briefly rinsed with isotonic saline by retrograde perfusion. Then, left ventricles were dissected, weighed, and homogenized on ice in protein extraction solution. The protocol has been optimized for mice cardiac ventricles weighing approximately 50 mg (*see* **Note 15**).

3.4.1 Sample Extraction and Nano-LC/MS Analysis

1. Homogenize the tissues in extraction solution keeping the ratio: 1:4 (w/v), tissue:solution.
2. Centrifuge the homogenate at 4 °C, 20,000 × *g* for 20 min.
3. Determine the total protein content of an aliquot of the homogenate using the BCA protein assay with bovine serum albumin as a standard.
4. Perform peptide extraction and nano-LC-MS/MS analysis as described previously in Subheadings 3.2–3.3.

Representative plasma angiotensin profile and cardiac angiotensin profile determined in 8-month-old male atherosclerotic ApoE$^{-/-}$/LDLR$^{-/-}$ and wild type C57BL/6J mice are presented on Figs. 3 and 4, respectively (*see* **Notes 16** and **17**).

Table 1

Molecular description and multiple reaction monitoring (MRM) settings for detection of angiotensin peptides

Peptide	Amino acid sequence	MW (Da)	Precursor ion (m/z)	Charge state	Product ion[a] (m/z)	Collision energy (eV)
Ang-(1–12)	DRVYIHPFHLLY	1572.8	524.92	3+	**583.58** ($b9^{2+}$) 109.65	18 33
Ang I (1–10)	DRVYIHPFHL	1296.5	432.91	3+	**647.19** ($b5^{+}$) 109.79	17 32
Ang-(1–9)	DRVYIHPFH	1183.3	395.10	3+	**156.03** (y^{+}) 109.85	20 33
Ang II (1–8)	DRVYIHPF	1046.2	523.85	2+	**784.26** ($b6^{+}$) 262.74	19 20
Ang A/ [Ala^1]-Ang II	ARVYIHPF	1002.2	501.62	2+	**740.36** ($b6^{+}$) 263.02	18 19
Ang-(1–7)	DRVYIHP	899.0	450.18	2+	**647.41** ($b5^{+}$) 109.97	19 28
Alamandine/ [Ala^1]-Ang-(1–7)	ARVYIHP	855.0	428.09	2+	**603.25** ($b5^{+}$) 109.77	17 23
Ang III (2–8)	RVYIHPF	931.1	466.19	2+	**669.46** ($b5^{+}$) 262.73	17 18
Ang IV (3–8)	VYIHPF	774.9	388.17	2+	**262.85** ($b2^{+}$) 136.07	12 20
[Asn^1, Val^5]-Ang II (IS)	NRVYVHPF	1031.2	516.19	2+	**769.34** ($b6^{+}$) 263.25	19 20

[a]Transition used for quantification of each peptide in biological sample refers to the transition in *bold*

4 Notes

1. Appropriate handling and subsequent extraction of plasma and tissue samples are critical to obtain reliable data on angiotensin levels, independent of the analytical method that is used for detection.
2. Regardless of the detection method used for angiotensin determination, such as RIA, RP-HPLC-RIA, or HPLC/MS,

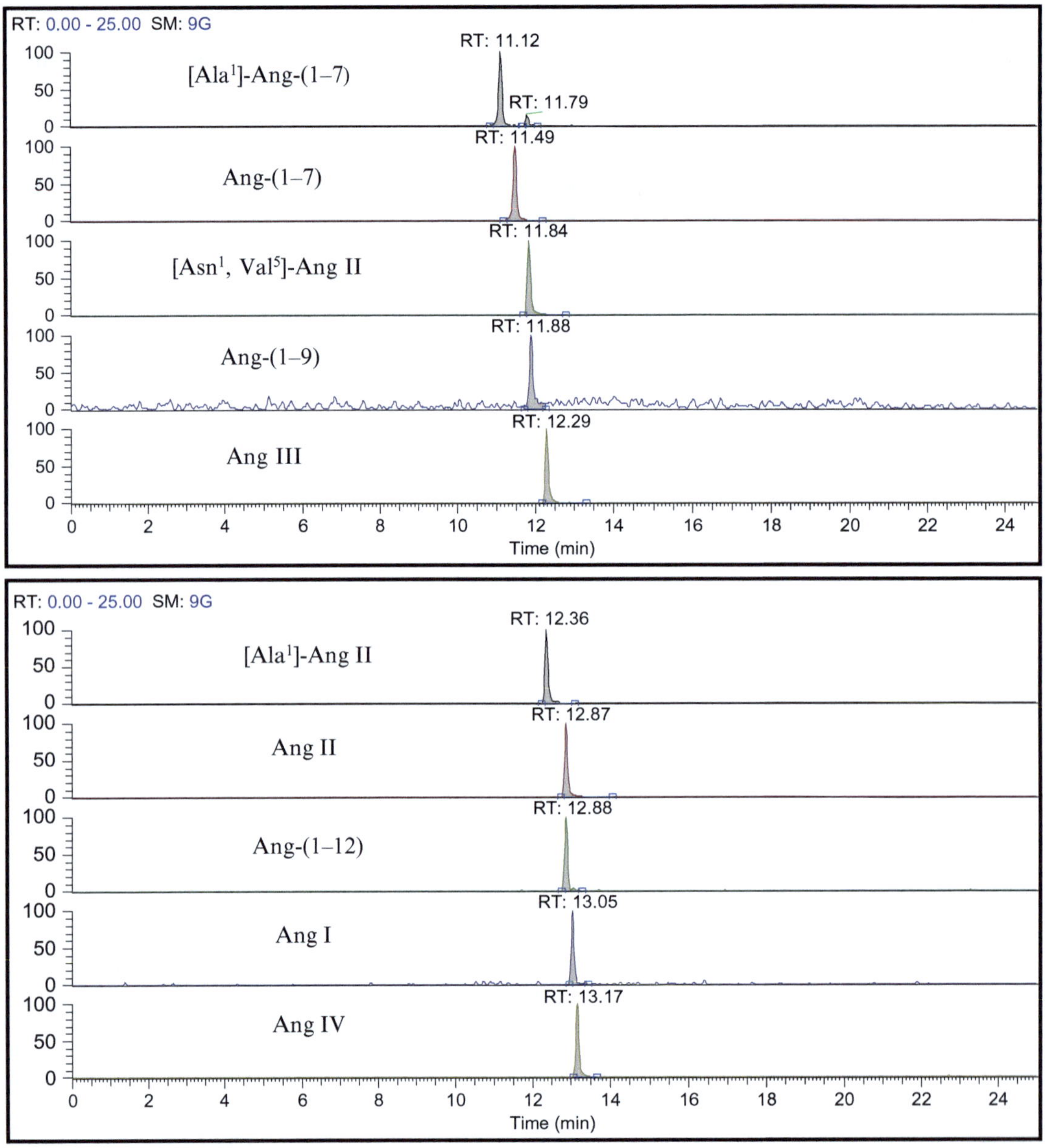

Fig. 2 Representative extract ion multiple reaction monitoring (MRM) chromatograms of nine angiotensin peptides and internal standard ([Asn^1, Val^5]-Ang II) obtained by analysis of a selected (ApoE$^{-/-}$/LDLR$^{-/-}$) unspiked plasma sample using nano-LC-MS/MS

improper sample handling and extraction may yield erroneously higher or lower concentrations.

3. Blood or tissue samples for angiotensin quantification should be immediately spiked with a potent protease inhibitor cocktail that enables efficient stabilization of angiotensin metabolites.

4. Because of above, the sampling time has to be kept as short as possible which could be challenging in some animal studies.
5. Optimal protease inhibitor cocktail blocking all angiotensin conversion enzymes has not been identified and still evolves with our expanding knowledge on peptide generation/degradation pathways.
6. Specific targets of each component of the Protease Inhibitor Cocktail are: EDTA and 1,10-phenanthroline act against metalloproteases; AEBSF acts to inhibit serine proteases, including trypsin, chymotrypsin-like enzymes, and plasmin among others; amastatin inhibits aminopeptidases, including aminopeptidase A; pepstatin A is an inhibitor of acid proteases (aspartyl peptidases) such as pepsin, renin, cathepsin D; thiorphan acts against neutral endopeptidase (NEP); and aliskiren is an inhibitor of renin activity.
7. Metalloenzymes such as ACE, ACE2, neprilysin (NEP), aminopeptidase A are the dominant class of peptidases and occurrence of divalent cations is essential for the preservation

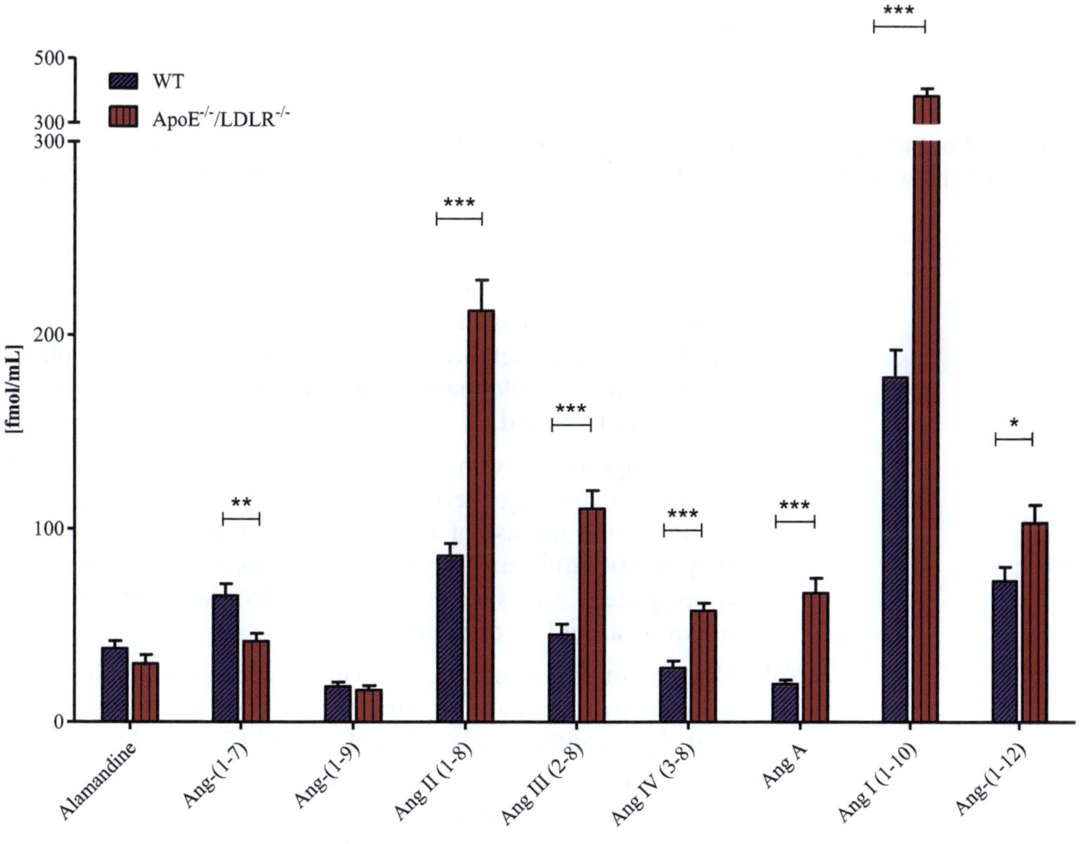

Fig. 3 Plasma angiotensin profile in 8-month-old WT (n = 5) and 8-month-old ApoE$^{-/-}$/LDLR$^{-/-}$ (n = 7) mice. Values were presented as means ± SEM. *P < 0.05, **P < 0.01, ***P < 0.001 vs. WT (age-matched) group

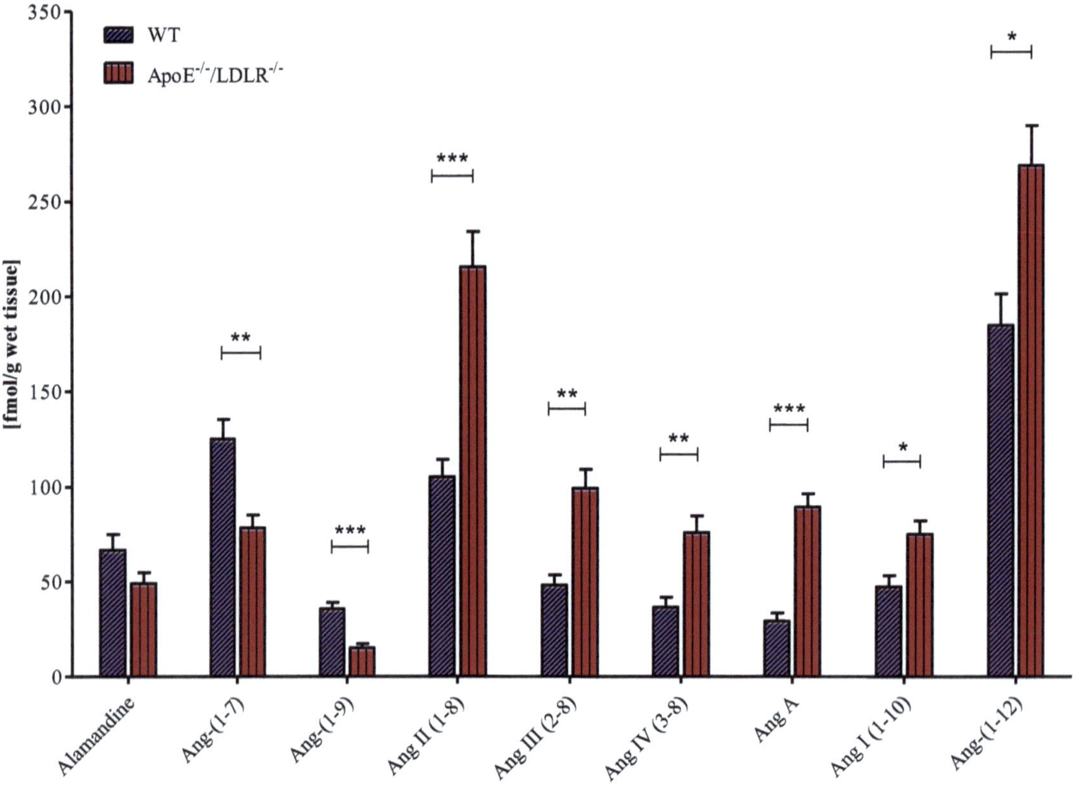

Fig. 4 Cardiac angiotensin profile in 8-month-old WT (n = 5) and 8-month-old ApoE$^{-/-}$/LDLR$^{-/-}$ (n = 7) mice. Values were presented as means ± SEM. *P < 0.05, **P < 0.01, ***P < 0.001 vs. WT (age-matched) group

of biological activity of these enzymes. However, strong chelating agents such as EDTA or phenanthroline, usually used to abolish peptide metabolism, are not sufficient to completely suppress the metalloenzyme activity and more specific inhibitors should be used.

8. When angiotensin–converting enzyme is inhibited, it leads to high renin and Ang I levels; speed of sample processing, quick freezing, and the use of a typical renin inhibitor – pepstatin A may not be sufficient to completely block renin activity, and a more potent and specific renin inhibitor such as enalkiren, aliskiren should be considered.
9. Before sample collection, animals should be maintained in a calm environment and anesthetic agents such as pentobarbital that significantly activate the RAS should be avoided.
10. Many earlier studies highlighted a significant loss of angiotensin peptides during storage, that was also noticed in the standard stock solutions [16, 18]. The reasons for this loss are not

clear but could be due to binding to storage vial walls. To avoid/reduce sample depletion by nonspecific peptide adsorption, only low-binding 0.5/1.5-mL vials and low-retention pipette tips were used. Moreover, fresh standards for each analysis series were used. Also, the samples were extracted and analyzed immediately to shorten the sample preparation time.

11. To optimize angiotensin recoveries and prevent irreversible adsorption of the peptides to the polypropylene tubes, the Tris-albumin-buffer coated tubes are often used [16, 21] but this procedure is more compatible with RIA-based assays.
12. So far, various extraction methods that block peptide metabolism, deplete the sample from contaminating proteins and other compounds that may interfere with the assay and enrich peptide content in a manageable volume have been developed [22]. They usually include protein precipitation (MeOH, HCl–EtOH, guanidine thiocyanate) and subsequent solid-phase extraction (C18 or phenyl). In our procedure, we have used acetonitrile precipitation followed by extraction on the C18 silica bonded SPE cartridges. Peptide recovery with our protocol for plasma samples typically ranges from 75 to 90%, whereas in tissues recoveries are generally lower in the 60–80% range.
13. RIA and ELISA are most common methods used to characterize angiotensins [22, 23]. Specificity of these methods is highly dependent on the characteristics for each antibody, as well as the proper sample extraction and assay conditions. The antibodies used in RIAs or ELISAs are typically directed against the unique COOH-terminus of the peptide, and distinction of the NH_2-terminal metabolites without additional chromatographic separation step can be problematic. LC/MS is therefore a more specific tool.
14. Our nano-LC-MS/MS approach enables to quantify comprehensively nine angiotensins in plasma and tissue samples with a sensitivity comparable to RIA and superior accuracy, high-throughput potential, and robustness.
15. Lower limits of quantification for angiotensin metabolites estimated by our nano-LC-MS/MS method are: 5 fmol/mL in plasma and 5 fmol/g in tissue samples.
16. The optimized nano-LC-MS/MS method allows the measurement of Ang metabolites in the same sample, providing a highly reliable estimate of their ratios. A specific metabolic pattern, called "RAS-fingerprint" could be drawn (Fig. 5) that provides visual information on the RAS alterations and possible changes in the angiotensin processing enzymes in the sample. This type of analysis may be well used for the monitoring of the efficacy of

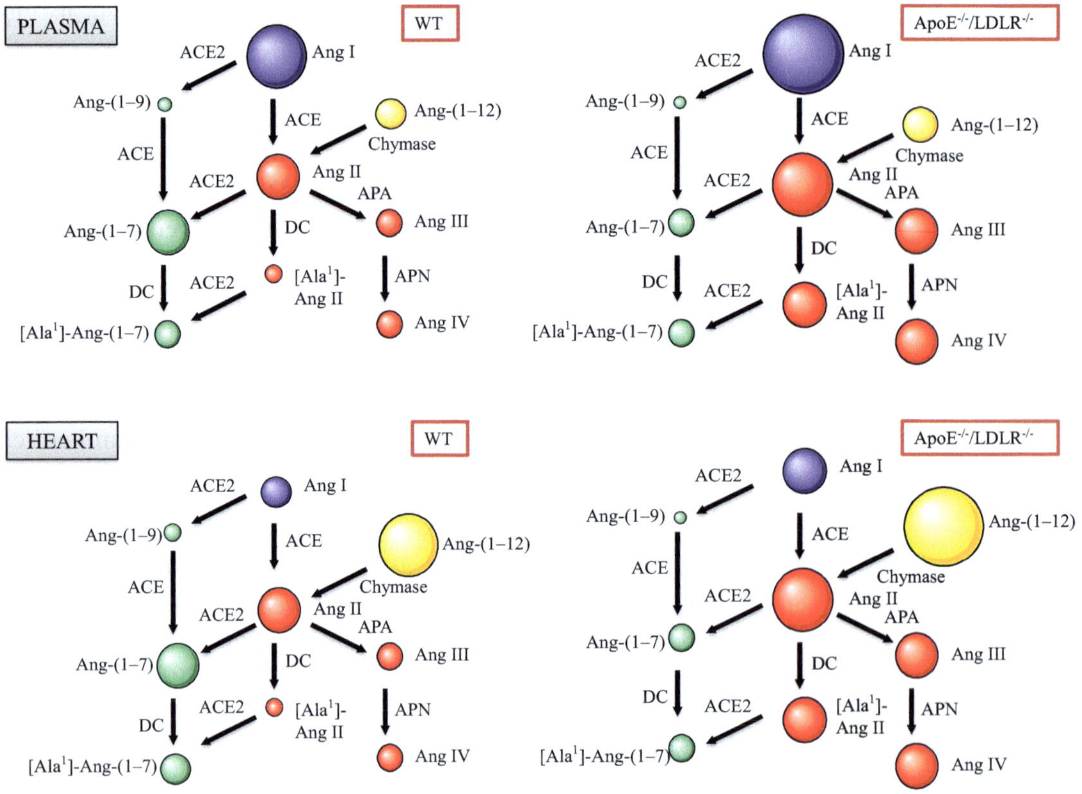

Fig. 5 Angiotensin peptide levels in WT and ApoE$^{-/-}$/LDLR$^{-/-}$ mice depicted using the diameter of the circles to reflect their concentration in plasma (upper panel) and heart (lower panel). The peptides formed by ACE2 cleavage are in very low concentrations (green) after the RAS dysregulation in atherosclerotic mice as opposed to those formed by ACE/chymase cleavage (red). Ang angiotensin, ACE angiotensin-converting enzyme, ACE2 angiotensin-converting enzyme 2, APA aminopeptidase A, APN aminopeptidase N, DC decarboxylase

the RAS-targeted therapies that constitute a major treatment modality in cardiovascular diseases [22, 23].

17. The accurate and highly sensitive procedure presented here combined with appropriate sampling conditions may bring our understanding of the RAS to the next level, although the results of nano-LC/MS and classical (RIA, ELISA) quantification methods still should be rigorously compared.

Acknowledgments

This work was supported by the European Union from the resources of the European Regional Development Fund under the Innovative Economy Programme (a grant coordinated by JCET-UJ, No. POIG.01.01.02-00-069/09) and TEAM programme of the Foundation for Polish Science (TEAM/2011-8/7).

References

1. Mehta PK, Griendling KK (2007) Angiotensin II cell signaling: physiological and pathological effects in the cardiovascular system. Am J Physiol Cell Physiol 292:C82–C97
2. Balakumar P, Jagadeesh G (2014) A century old renin-angiotensin system still grows with endless possibilities: AT1 receptor signaling cascades in cardiovascular physiopathology. Cell Signal 26:2147–2160
3. Te Riet L, van Esch JH, Roks AJ, van den Meiracker AH, Danser AH (2015) Hypertension: renin-angiotensin-aldosterone system alterations. Circ Res 116:960–975
4. Oparil S, Schmieder RE (2015) New approaches in the treatment of hypertension. Circ Res 116:1074–1095
5. Passos-Silva DG, Brandan E, Santos RA (2015) Angiotensins as therapeutic targets beyond heart disease. Trends Pharmacol Sci 36:310–320
6. Donoghue M, Hsieh F, Baronas E et al (2000) A novel angiotensin-converting enzyme-related carboxypeptidase (ACE2) converts angiotensin I to angiotensin 1-9. Circ Res 87:E1–E9
7. Keidar S, Kaplan M, Gamliel-Lazarovich A (2007) ACE2 of the heart: from angiotensin I to angiotensin (1–7). Cardiovasc Res 73:463–469
8. McKinney CA, Fattah C, Loughrey CM, Milligan G, Nicklin SA (2014) Angiotensin-(1–7) and angiotensin-(1–9): function in cardiac and vascular remodelling. Clin Sci (Lond) 126:815–827
9. Jiang F, Yang J, Zhang Y et al (2014) Angiotensin-converting enzyme 2 and angiotensin 1–7: novel therapeutic targets. Nat Rev Cardiol 11:413–426
10. Jankowski V, Vanholder R, van der Giet M et al (2007) Mass-spectrometric identification of a novel angiotensin peptide in human plasma. Arterioscler Thromb Vasc Biol 27:297–302
11. Lautner RQ, Villela DC, Fraga-Silva RA et al (2013) Discovery and characterization of alamandine: a novel component of the renin-angiotensin system. Circ Res 112:1104–1111
12. Etelvino GM, Peluso AA, Santos RA (2014) New components of the renin-angiotensin system: alamandine and the MAS-related G protein-coupled receptor D. Curr Hypertens Rep 16:433
13. Ferrario CM (2010) New physiological concepts of the renin-angiotensin system from the investigation of precursors and products of angiotensin I metabolism. Hypertension 55:445–452
14. Allred AJ, Chappell MC, Ferrario CM, Diz DI (2000) Differential actions of renal ischemic injury on the intrarenal angiotensin system. Am J Physiol Renal Physiol 279:F636–F645
15. Mazzolai L, Pedrazzini T, Nicoud F, Gabbiani G, Brunner HR, Nussberger J (2000) Increased cardiac angiotensin II levels induce right and left ventricular hypertrophy in normotensive mice. Hypertension 35:985–991
16. Nussberger J, Brunner DB, Nyfeler JA, Linder L, Brunner HR (2001) Measurement of immunoreactive angiotensin-(1–7) heptapeptide in human blood. Clin Chem 47:726–729
17. Yamamoto K, Ohishi M, Katsuya T et al (2006) Deletion of angiotensin-converting enzyme 2 accelerates pressure overload-induced cardiac dysfunction by increasing local angiotensin II. Hypertension 47:718–726
18. Lortie M, Bark S, Blantz R, Hook V (2009) Detecting low-abundance vasoactive peptides in plasma: progress toward absolute quantitation using nano liquid chromatography-mass spectrometry. Anal Biochem 394:164–170
19. Olkowicz M, Radulska A, Suraj J et al (2015) Development of a sensitive, accurate and robust liquid chromatography/mass spectrometric method for profiling of angiotensin peptides in plasma and its application for atherosclerotic mice. J Chromatogr A 1393:37–46
20. Olkowicz M, Chlopicki S, Smolenski RT (2015) Perspectives for angiotensin profiling with liquid chromatography/mass spectrometry to evaluate ACE/ACE2 balance in endothelial dysfunction and vascular pathologies. Pharmacol Rep 67:778–785
21. Nussberger J, Brunner DB, Waeber B, Brunner HR (1985) True versus immunoreactive angiotensin II in human plasma. Hypertension 7:I1–I7
22. Chappell MC (2016) Biochemical evaluation of the renin-angiotensin system: the good, bad, and absolute? Am J Physiol Heart Circ Physiol 310:H137–H152
23. Poglitsch M, Sturrock ED, Danser AH (2016) Letter to the editor: angiotensin quantification by mass spectrometry. Am J Physiol Heart Circ Physiol 310(33):H452–H453

Chapter 15

Analysis of Angiotensin Metabolism in the Kidney Using Mass Spectrometry

Nadja Grobe and Khalid M. Elased

Abstract

The renin angiotensin system (RAS) is a highly complex enzymatic system consisting of multiple peptide hormones, enzymes, and receptors. A thorough characterization of angiotensin peptide metabolism is crucial for understanding pathological states associated with an imbalanced RAS. Here, we describe two matrix-assisted laser desorption/ionization (MALDI) mass spectrometric (MS) approaches for the assessment of in vitro and in situ RAS enzymatic activities in the kidney using the natural angiotensin peptide substrates. These MS techniques demonstrate high specificity and are superior over conventional spectrophotometric or colorimetric assays since multiple proteolytic cleavage sites can be detected, thus unraveling the complexity of the RAS.

Key words Mass spectrometry, MALDI, Imaging, Renin-angiotensin system, Kidney, Enzyme activity, Angiotensin metabolism, ACE, ACE2, NEP

1 Introduction

The renin-angiotensin system (RAS) plays a fundamental role in the pathophysiology of hypertension and the onset and progression of renal injury. However, the intrarenal RAS network and the role of specific peptidases in mediating renal and cardiovascular health are incompletely understood partly due to limited analytical techniques that are able to unravel its complexity. Mass spectrometry (MS) is a fundamental technique that has been widely used to characterize cell systems by assessing metabolites, peptides, and proteins. Over the last 15 years, several MS approaches have been developed to measure the RAS, replacing traditional fluorogenic substrate, colorimetric, ultraviolet, or radioimmunoassay (RIA) based techniques. Since molecules are ionized and analyzed according to their mass-to-charge (m/z) ratios, the natural angiotensin (Ang) peptide substrates and products resulting from RAS peptidase cleavage are more selectively and specifically detected using MS. Indeed, Fredline et al. showed for the first time a higher

Sean E. Thatcher (ed.), *The Renin-Angiotensin-Aldosterone System: Methods and Protocols*, Methods in Molecular Biology, vol. 1614, DOI 10.1007/978-1-4939-7030-8_15, © Springer Science+Business Media LLC 2017

specificity for assessing plasma renin activity using MS as compared to RIA [1]. MS was also used to characterize various substrates and multiple proteolytic cleavage sites for the newly discovered angiotensin converting enzyme 2 (ACE2), a novel component of the protective arm of the RAS [2]. We and others developed several MS-based assays to detect and quantify metabolism of the RAS peptides in cells, plasma, tissue homogenates, and tissue sections [3–8]. Although a local RAS exists in many organ systems, we have extensively focused our studies on the in vitro and in situ metabolism of the intrarenal RAS due to its pivotal role in blood pressure regulation, salt and acid–base homeostasis, and pathogenesis of kidney diseases. Compared to other ionization techniques and liquid chromatography-based approaches, matrix-assisted laser desorption/ionization (MALDI) MS is uniquely suited for the characterization of RAS peptidase activities because of a softer ionization technique, easy-to-follow sample preparation, short analysis time, and high-throughput capability. Here, we demonstrate that the intrarenal RAS peptidases and metabolism of several

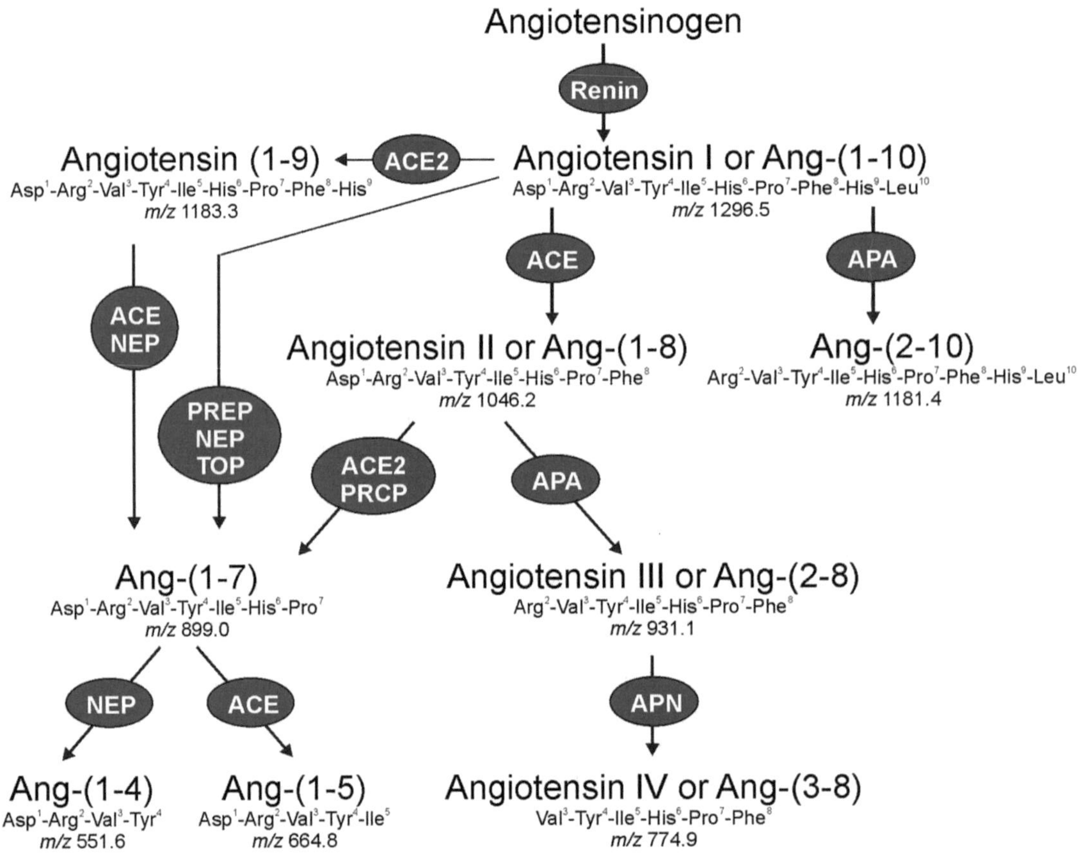

Fig. 1 Angiotensin metabolism in the kidney detected by mass spectrometry. *ACE* Ang converting enzyme, *NEP* neprilysin, *ACE2* Ang converting enzyme 2, *APA* aminopeptidase A, *APN* aminopeptidase N, *PRCP* prolyl carboxypeptidase, *PREP* prolyl endopeptidase, *TOP* thimet oligopeptidase

Ang peptides can be efficiently measured using MALDI MS. Our current RAS test platform is capable of localizing the multistep enzymatic cascade of the RAS, starting from the decapeptide Ang I down to smaller Ang peptides such as Ang-(1-4) (Fig. 1).

2 Materials

Prepare all solutions using ultrapure, LC-MS grade water and LC-MS grade reagents. Prepare and store all reagents at room temperature (unless indicated otherwise). Diligently follow all waste disposal regulations when disposing of waste materials.

2.1 MS Analysis of In Vitro Renal Angiotensin Metabolism

1. Phenylmethanesulfonyl flouride (PMSF): 100 mM PMSF in dimethyl sulfoxide (DMSO). Store at −20 °C.
2. Lysis buffer: Dissolve one tablet protease inhibitor in 10 ml Complete Lysis-M EDTA-free lysis buffer and add 250 μl 100 mM PMSF. Store at 4 °C.
3. Kimble Kontes Duall Tissue Grinder, size 22.
4. Bovine serum albumin (BSA) standard for protein assay: Prepare 1 mg BSA in 1 ml water. Store at −20 °C.
5. 2-(N-morpholino)ethanesulfonic acid (MES) buffer: 0.4 M MES buffer, pH 6.75. Weigh 3.9 g MES hydrate and add about 30 ml of water to a 100-ml beaker. Adjust pH using 1 N NaOH. Make up to 50 ml with water.
6. Bestatin: 500 μM in 100% isopropyl alcohol. Store at −20 °C.
7. Ang peptides: Prepare 1 mg Ang peptide in 1 ml water. Store at −20 °C.
8. RAS inhibitors: Prepare 10 mM in water (*see* **Note 1**). Store at −20 °C.
9. Stop solution: 10% trifluoroacetic acid (TFA) in water.
10. Diluent: 90% acetonitrile–0.3% TFA (v/v). Mix 900 μl acetonitrile with 30 μl 10% TFA and 70 μl water.
11. Ang internal standards: stable-isotope, [$^{13}C_5$, ^{15}N]-Valine labeled Ang peptide standards aliquoted as 1000 pmol/μl in 30% acetonitrile with 0.1% formic acid and stored at −80 °C. Dilute 1:10 with water (*see* **Note 2**).
12. MALDI matrix: 10 mg/ml α-cyano-4-hydroxycinnamic acid (CHCA), 60% methanol, 10% acetone, and 0.3% TFA (*see* **Note 3**). Dissolve 10 mg α-cyano-4-hydroxycinnamic acid in 0.6 ml methanol, 0.1 ml acetone, 0.03 ml 10% TFA, and 0.27 ml water.
13. Peptide calibration standard II: Peptide mixture for calibration of matrix-assisted laser desorption and ionization time-of-flight

(MALDI-TOF) mass spectrometer (MS) consisting of the peptides Bradykinin 1–7, Angiotensin II, Angiotensin I, Substance P, Bombesin, ACTH clip 1–17, ACTH clip 18–39, and Somatostatin 28. Dissolve one tube in 125 μl 0.1% TFA and store at −20 °C. Spot 0.5 μl peptide calibration standard II and mix with 0.5 μl MALDI matrix. Spot peptide calibration standard II onto each corner of the 384 ground steel MALDI target plate (*see* **Note 4**).

2.2 MS Analysis of In Situ Renal Angiotensin Metabolism

1. Ang peptides and RAS inhibitors.
2. Peptide calibration standard II. Spot onto indium-tin-oxide coated glass slides.
3. MALDI matrix: prepare 20 ml of 10 mg/ml α-cyano-4-hydroxycinnamic acid, 60% methanol, 10% acetone, and 0.3% TFA. Dissolve 200 mg CHCA in 12 ml methanol, 2 ml acetone, 0.6 ml 10% TFA, and 5.4 ml of water (*see* **Note 3**).
4. Slide moisture chamber.
5. Laboratory incubator set to 37 °C.
6. Vacuum desiccator connected to a vacuum pump.
7. Scanner.
8. 10-ml thin layer chromatography sprayer connected to a Drierite gas drying unit and nitrogen gas cylinder.
9. Light microscope with 10× magnification.

3 Methods

3.1 MS Analysis of In Vitro Renal Angiotensin Metabolism

1. Kidney collection: Quickly remove the kidney, transfer into a 1.5-ml Eppendorf tube, and immediately freeze using liquid nitrogen or dry ice. Store at −80 °C.
2. Kidney homogenates: Rinse mouse kidney twice in 1 ml ice-cold Roche EDTA-free protease inhibitor buffer containing 2.5 mM PMSF. Transfer washed kidney to a chilled glass homogenizer and homogenize on ice using 30 strokes. Transfer first homogenate to a chilled 5-ml tube. Add 1 ml to the chilled homogenizer and homogenize remaining kidney pieces with five strokes. Combine second homogenate with first homogenate in 5-ml tube. Divide combined homogenate into two chilled 1.5-ml tubes and centrifuge at 9279 × *g* for 10 min and 4 °C. Combine supernatants in a fresh, chilled 5-ml tube and prepare 100-μl aliquots to be stored at −80 °C.
3. Protein assay: Determine protein content of kidney homogenate using BSA standard and Bradford reagent according to the manufacturer's instructions.

4. Incubation: Mix 20 µg kidney homogenate protein with 12.5 µl 0.4 M MES buffer, pH 6.75, and 2 µl 100 mM PMSF in a 1.5 ml tube (*see* **Note 5**). Make up to 100 µl water. Start reaction by adding 5 µl 1 mg/ml Ang peptide (*see* **Note 6**). Incubate for 30–120 min at 37 °C in a thermomixer under shaking at 350 rotations per minute (rpm).
5. Inhibitor studies: Prepare the same reaction mixture as in Subheading 3.1, **step 4**, and add 1 µl 10 mM inhibitor (*see* **Note 7**). Make volume up to 100 µl water and pre-incubate test tubes for 10 min at 37 °C at 350 rpm. Start reaction by adding Ang peptide. Incubate for 30–120 min at 37 °C in a thermomixer under shaking at 350 rpm.
6. MS sample preparation: Add 10 µl 10% TFA and 10 µl 1:10 diluted stable isotope labeled Ang internal standards to reaction mixture. Take out 2 µl and mix with 18 µl of diluent consisting of 90% acetonitrile containing 0.3% TFA, vortex, centrifuge, spot 0.5 µl onto a 384 MALDI target plate and mix with 0.5 µl MALDI matrix.
7. MS acquisition: Mass calibrate by collecting 200 laser shots of Bruker peptide calibration standard II using an Autoflex III smartbeam MALDI TOF/TOF instrument (Bruker Daltonics, MA, USA). Operate mass spectrometer with positive polarity in reflectron mode and analyze spots using the method "RP-pepmix." Acquire a total of 2000–3000 laser shots (2–3× 1000 shots) randomly for each spot in the range of *m/z* 500–3000 at a laser frequency of 100 Hz. Identify and confirm detected molecules using the Bruker Lift method upon comparison to standard compounds (Fig. 2).
8. Data analysis: Analyze data with proprietary Bruker Flex Analysis software using baseline subtraction, smoothing, and normalization. Quantitate MALDI signals for Ang peptides by comparing to signal intensities obtained for the stable-isotope labeled Ang internal standards (Fig. 3).

3.2 MS Analysis of In Situ Renal Angiotensin Metabolism

1. Kidneys.
2. Kidney sections: Let tissue adjust to −20 °C for 30 min. Cut kidney sagittally in half using a chilled blade. Cut 12 µm-sized tissue sections at −20 °C using a cryostat (*see* **Note 8**). Move tissue sections carefully on indium-tin-oxide coated glass slides at −20 °C using chilled brushes (*see* **Note 9**). Mount tissue section onto glass slide by holding finger underneath the glass slide containing the tissue section for 1 min at −20 °C. Desiccate tissue sections for 5 min at room temperature.
3. Incubation: Apply 5 µl of 1 mg/ml Ang peptide to tissue sections using a 20-µl pipet at room temperature. For inhibitor studies, add 1 µl 10 mM inhibitor to 100 µl 1 mg/ml Ang

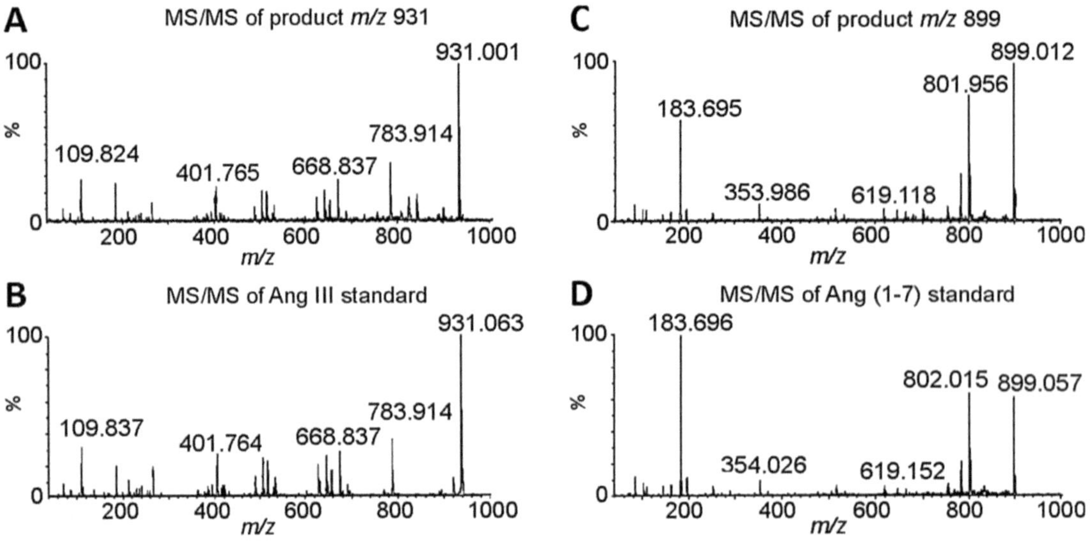

Fig. 2 Confirmation of detected Ang III and Ang-(1-7) peptides upon comparison to standard compounds using the Bruker Lift method. (**a**) Shows the MS/MS of Ang III (*m/z* 931) produced by kidney homogenate after incubation with Ang II. (**b**) Shows the MS/MS of the peptide standard Ang III (*m/z* 931). (**c**) Shows the MS/MS of Ang-(1-7) (*m/z* 899) produced by kidney homogenate after incubation with Ang II. (**d**) Shows the MS/MS of the peptide standard Ang-(1-7) (*m/z* 899)

peptide solution (*see* **Note 7**). Incubate tissue sections with test solution in humidity chamber placed inside an incubator at 37 °C for 5 min (*see* **Note 10**).

4. Preparing tissue sections and glass slides for matrix coating: Desiccate tissue sections at room temperature for at least 30 min. When dry, spot immediately 0.5 μl peptide calibration standard II next to the tissue sections in a corner of the indium-tin-oxide coated glass slide and mix with 0.5 μl MALDI matrix. Let spot dry at room temperature for 5 min. Add teaching points to each corner of the indium-tin-oxide coated glass slide using a permanent marker and white-out. Scan glass slide containing tissue sections, peptide calibration standard, and teaching points.
5. Matrix coating of tissue sections: Spray 20 ml MALDI matrix per glass slide using a thin layer chromatography nebulizer with nitrogen gas (10 psi) from a distance of 20 cm. Repeat spray coating 10–15× allowing for 30 s of drying time between 3 and 5 passes until a uniform matrix coating is achieved. Desiccate tissue sections at room temperature for at least 10 min. Use light microscopic visualization at 10× magnification to confirm even matrix coating.
6. MS acquisition: Mass calibrate spectra by collecting 200 laser shots of Bruker peptide calibration standard II. Start automated analysis of tissue sections using an Autoflex III

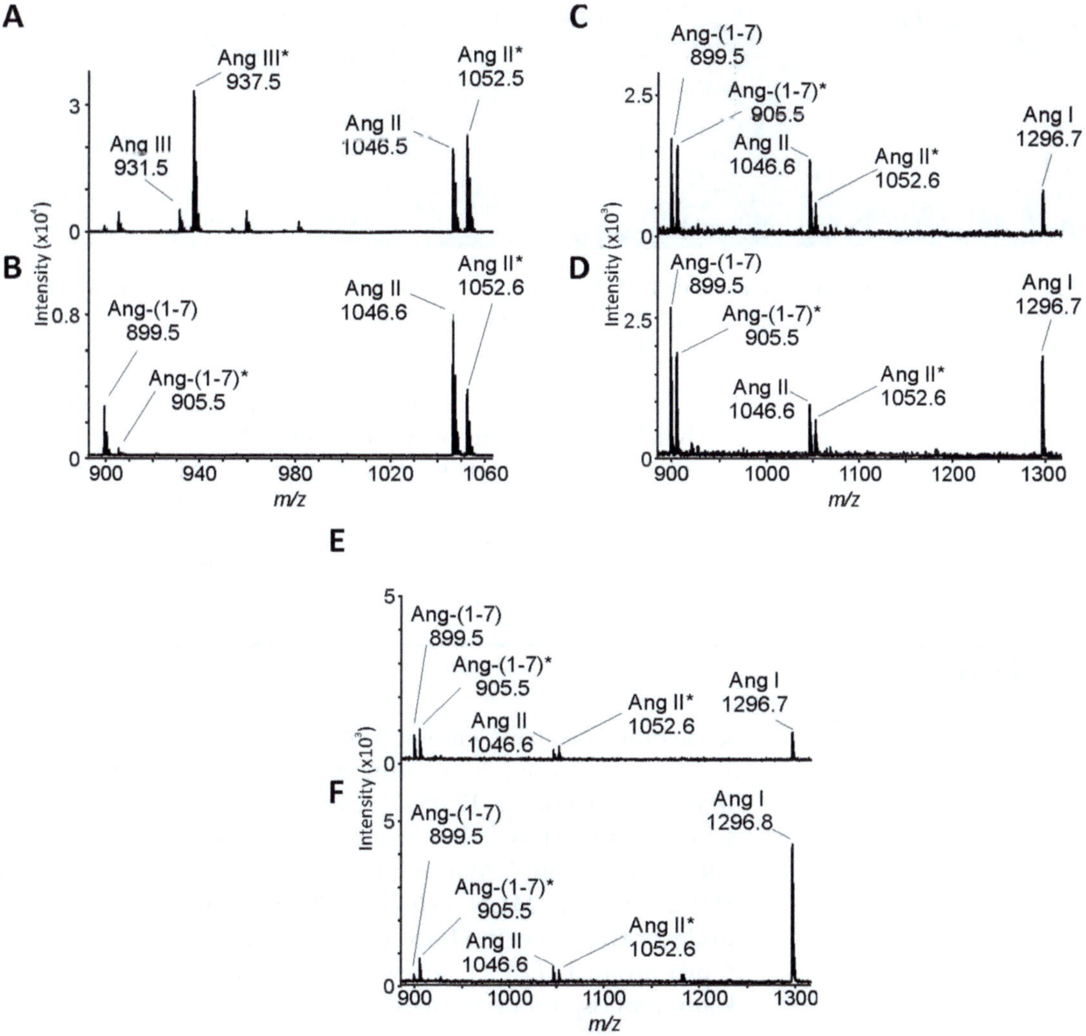

Fig. 3 MS analysis of in vitro renal angiotensin metabolism. (**a**) Shows the MS signals obtained after incubation of Ang II with 20 μg kidney homogenate protein in 50 mM phosphate buffer pH 7.0 for 60 min. (**b**) Shows the MS signals obtained after incubation of Ang II with 20 μg kidney homogenate protein in 50 mM citrate buffer pH 5.0 for 60 min. (**c**, **d**) Show the MS signals obtained after 60-min incubation of Ang I with 20 μg kidney homogenate protein in 50 mM MES buffer, pH 6.75, in absence or presence of 100 μM ACE inhibitor captopril, respectively. (**e**, **f**) Show the MS signals obtained after a 60-min incubation of Ang I with 20 μg kidney homogenate protein in 50 mM MES buffer pH 6.75 in absence or presence of 100 μM NEP inhibitor thiorphan, respectively. The *asterisk* indicates stable-isotope, [$^{13}C_5$, ^{15}N]-Valine labeled Ang peptide standards

smartbeam MALDI TOF/TOF instrument (Bruker Daltonics, MA, USA) in positive polarity and reflectron mode acquiring a total of 200 spectra for each spot of a 100 or 200 μm raster in the range of *m/z* 500–3000 and at a laser frequency of 100 Hz. Identify and confirm detected molecules using the Bruker Lift method upon comparison to standard compounds.

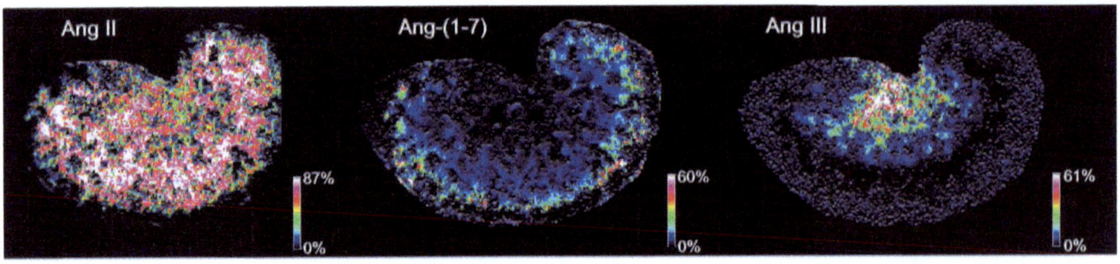

Fig. 4 MS analysis of in situ renal angiotensin metabolism. Incubation of mouse kidney sections with Ang II shows distinct spatial distribution patterns for the two main products Ang-(1-7) and Ang III

7. Data analysis: Analyze date with proprietary Bruker Flex Analysis and Imaging software using baseline subtraction, smoothing, and normalization. Quantitate MALDI imaging signals represented in color on pseudocolor images (Fig. 4) as integrated intensity and normalized to the area of the whole kidney section using MetaMorph image analysis software.

4 Notes

1. Some inhibitors have a low solubility in water and need sonication to be dissolved. Some inhibitors need a few initial microliters of ethanol or DMSO. Check with manufacturer for detailed instructions. For instance, we found the following solubilities: 10 mM ACE inhibitor captopril in water, 10 mM ACE2 inhibitor MLN-4760 (MLN) in water, 10 mM NEP inhibitor thiorphan in DMSO, 10 mM PRCP/PREP inhibitor Z-pro-prolinal (ZPP) in water, 10 mM APA inhibitor 4-aminophosphobutyric acid in water.
2. Prepare the dilution of Ang internal standards freshly right before use.
3. Prepare MALDI matrix freshly right before use. Vortex thoroughly.
4. Clean target plate before first use with isopropanol followed by water. Wet a tissue and wipe surface. Sonicate the target plate for 10 min covered with isopropanol and dry the target plate using a stream of nitrogen or let it air dry in a dust-free environment. The target plate can be reused following the cleaning procedure.
5. Buffers other than MES buffer, pH 6.75, may be used to characterize the RAS. We have used 50 mM phosphate buffer, pH 7, or 50 mM citrate buffer, pH 5 (Fig. 3).
6. If Ang II is tested as substrate, add 4 μl 500 μM Bestatin before making up to 100 μl water to stabilize Ang II.
7. Different inhibitors can be used to test RAS enzyme activities, including ACE inhibitor captopril, ACE2 inhibitor MLN-

4760, NEP inhibitor thiorphan, PRCP/PREP inhibitor Z-pro-prolinal, or APA inhibitor 4-aminophosphobutyric acid. Figure 3 shows an example of kidney homogenate incubated with Ang I in absence and presence of ACE inhibitor captopril or NEP inhibitor thiorphan.

8. To obtain intact kidney sections, adjust temperature of the cryostat to −20 ± 3 °C. Cutting performance depends on room temperature and humidity. Cryostat temperature has to be adjusted accordingly.
9. One indium-tin-oxide coated glass slide can hold up to 12 mouse kidney sections.
10. Subheading 3.2, **steps 1–5** are key features unique to this technique. These steps are highly temperature- and time-sensitive. Proper processing is crucial for visualization of tissue enzyme activities, including (a) quick removal and rapid freezing of organs after collection, (b) avoidance of localized tissue warming and contamination of tissue cutting surface with freezing medium, (c) care in applying the peptide solution such that the tissue is not damaged, and (d) minimal movement of slides during incubation and transfer to desiccator.

Acknowledgment

This work was supported by National Institutes of Health Grants F32 DK-093226 and the Carl W. Gottschalk Research Scholar Grant of the ASN Foundation for Kidney Research to N.G. and R01 HL-093567 to K.M.E.

References

1. Fredline VF, Kovacs EM, Taylor PJ, Johnson AG (1999) Measurement of plasma renin activity with use of HPLC-electrospray-tandem mass spectrometry. Clin Chem 45:659–664
2. Donoghue M, Hsieh F, Baronas E, Godbout K, Gosselin M, Stagliano N, Donovan M, Woolf B, Robison K, Jeyaseelan R, Breitbart RE, Acton S (2000) A novel angiotensin-converting enzyme-related carboxypeptidase (ACE2) converts angiotensin I to angiotensin 1-9. Circ Res 87:E1–E9
3. Elased KM, Cool DR, Morris M (2005) Novel mass spectrometric methods for evaluation of plasma angiotensin converting enzyme 1 and renin activity. Hypertension 46:953–959
4. Elased KM, Cunha TS, Gurley SB, Coffman TM, Morris M (2006) New mass spectrometric assay for angiotensin-converting enzyme 2 activity. Hypertension 47:1010–1017
5. Grobe N, Elased KM, Cool DR, Morris M (2012) Mass spectrometry for the molecular imaging of angiotensin metabolism in kidney. Am J Physiol Endocrinol Metab 302: E1016–E1024
6. Grobe N, Di FM, Kashkari N, Chodavarapu H, Somineni HK, Singh R, Elased KM (2015) Functional and molecular evidence for expression of the renin angiotensin system and ADAM17-mediated ACE2 shedding in COS7 cells. Am J Physiol Cell Physiol 308: C767–C777
7. Stettner D, Bujak-Gizycka B, Olszanecki R, Rytlewski K, Huras H, Korbut R (2013) Assessment of angiotensin I metabolism in the human placenta using an LC/MS method. Folia Med Cracov 53:31–39
8. Velez JC, Ierardi JL, Bland AM, Morinelli TA, Arthur JM, Raymond JR, Janech MG (2012) Enzymatic processing of angiotensin peptides by human glomerular endothelial cells. Am J Physiol Renal Physiol 302:F1583–F1594

ERRATUM TO

Chapter 8
Characterization and Functional Phenotyping of Renal Immune Cells via Flow Cytometry

Nathan P. Rudemiller and Steven D. Crowley

Sean E. Thatcher (ed.), *The Renin-Angiotensin-Aldosterone System: Methods and Protocols*, Methods in Molecular Biology, vol. 1614, DOI 10.1007/978-1-4939-7030-8_8, © Springer Science+Business Media LLC 2017

DOI 10.1007/978-1-4939-7030-8_16

This chapter was originally published with two errors in the Materials section found on pages 88 and 89.

14. should read "…reconstitute in PBS at **100** mg/ml…" not 0.1 mg/ml as is written

16. should read "…containing 3% FBS, 10 **ug**/ml DNase…" not 10 mg/ml as is written

The updated original online version for this chapter can be found at http://dx.doi.org/10.1007/978-1-4939-7030-8_8

INDEX

Sean E. Thatcher (ed.), *The Renin-Angiotensin-Aldosterone System: Methods and Protocols*, Methods in Molecular Biology, vol. 1614, DOI 10.1007/978-1-4939-7030-8, © Springer Science+Business Media LLC 2017

MIX
Papier aus verantwortungsvollen Quellen
Paper from responsible sources
FSC® C105338

If you have any concerns about our products,
you can contact us on
ProductSafety@springernature.com

In case Publisher is established outside the EU,
the EU authorized representative is:
Springer Nature Customer Service Center GmbH
Europaplatz 3, 69115 Heidelberg, Germany

Printed by Libri Plureos GmbH
in Hamburg, Germany